Redesigning Life

Eugenics, Biopolitics, and the Challenge of the Techno-Human Condition

P.I.E. Peter Lang

Bruxelles · Bern · Berlin · Frankfurt am Main · New York · Oxford · Wien

Nathan Van Camp

Redesigning Life

Eugenics, Biopolitics, and the Challenge of the Techno-Human Condition

Philosophy & Politics
No.27

Published with the support of the University Foundation of Belgium

This publication has been peer-reviewed.

1 avenue Maurice, B-1050 Bruxelles, Belgique
www.peterlang.com ; info@peterlang.com

ISSN 1376-0920
ISBN 978-2-87574-281-0
eISBN 978-3-0352-6544-6
D/2015/5678/40

CIP available from the Library of Congress USA, and the British Library, GB.

Bibliographic information published by "Die Deutsche Bibliothek"
"Die Deutsche Bibliothek" lists this publication in the "Deutsche Nationalbibliografie"; detailed bibliographic data is available in the Internet at <http://dnb.de>.

Table of Contents

Acknowledgments

Some of the work presented in this book has already appeared in academic journals. For material that is presented in Chapters 1, 3, 4, and 5, the author acknowledges its original publication in a slightly different form in *Humana Mente: Journal of Philosophical Studies*, *Journal of Philosophy of Life*, *Symposium: Canadian Journal of Continental Philosophy*, *Between the Species*, and *Journal for Cultural Research*. The author sincerely thanks the editors of these journals for their permission to publish this work.

I am grateful to Gabriel Fragnière and Hendrik Opdebeeck, the series editors for Philosophy & Politics at Peter Lang, for publishing this book. My thanks go out, as well, to the managing director at PIE Peter Lang, Emilie Menz, for her patience, thoroughness, and belief in the project.

This book is derived from my doctoral dissertation which was submitted to and successfully defended at the Department of Philosophy of the University of Antwerp (Belgium) in May 2013. A project like this one requires assistance from many quarters. I give special thanks to my parents, Marc Van Camp and Anita De Hoef, for their unconditional support throughout my extended stay at the University of Antwerp. I am profoundly grateful. I also thank my brothers, grandparents, and family-in-law. They were always there when I needed them, and I am truly grateful for that.

Others whom I want to acknowledge here are all good friends who I have known for many years and who have given me the courage and strength to pursue my dreams: Bert De Smedt, Lente Dewilde, Joris Godefroit, Ines Borgonjon, Pieter Thomaes, Stephanie Williams, Nicolas Dobbeleir, Stefan Engelen, Stien Smessaert, Davy Rombouts, Kathelijne Delmeire, Dennis Soete, Griet Debussche, Winnok De Vos, Katrien Van Den Broeck, Tim Dupuis, Laïcha De Cock, Steve Leung, Maria Salinas Sanchez, Carlos Rovzar, Pieter Vereycken, Sarah Shaham, Annelies Godefroit, Sophie Stas, Christophe Huybrighs, and many others who I have forgotten to mention here. I am honored to call them my friends. I also want to thank a couple of people who have come to me from the world of academia and who have each, in their own way, contributed to the coming into being of this book: Diane Enns, Jeffrey A. Barash, Bart Verschaffel, Walter Weyns, Kris Dierickx, Francis D. Raška, Cassie Striblen, Stephen Bloch-Schulman, Sophie Cloutier, Joshua Miller, Shane Wilkins, Eli Schonfeld, and Paula Schwebel.

For the generous scholarship that allowed me to carry out the research that led to this book, I express my gratitude to the Research Foundation Flanders (FWO). I also thank the University Foundation of Belgium for its financial support in making this publication possible.

I am profoundly grateful to my colleagues and former colleagues in the Institute of Jewish Studies as well as in the philosophy department at the University of Antwerp. In particular, I acknowledge Joachim Leilich, Katrien Vloeberghs, Karolien Vermeulen, Luc Acke, Christophe Collard, Jo Bogaerts, Dennis Baert, David Dessin, Liesbet Quaeghebeur, Sven Braspenning, Jeff Spiessens, Geoffrey Dierickxens, Rudmer Bijlsma, Martin Michels, and Torben Wolfs.

Finally, I thank two mentors, without whom this whole enterprise would have been unattainable. Arthur Cools, chair of the Center for European Philosophy at the University of Antwerp, first sparked my interest in the broad field of what is now often referred to as "continental philosophy." For this, and for his support and critical guidance over the years, I owe him a debt of gratitude. Last, but not least, I am forever grateful to my *Doktormutter*, Vivian Liska, director of the Institute of Jewish Studies at the University of Antwerp, for her guidance and encouragement, not only during my doctoral studies, but ever since. Without her steadfast belief in my potential as a scholar, I would simply not be the person who I am today. There are not enough words in my mind to thank her properly.

I have dedicated this book to the love of my life, Leen. This book would simply not have been possible without her unwavering love, support, patience, and belief in me as a scholar and as a person. Our love journey started well before I embarked on the journey to wisdom, and my life as a philosopher would simply be meaningless without her love. I am truly grateful in ways that I may never be able to show.

Introduction

In the late 1990s, a team of Princeton geneticists led by Joe Tsien succeeded in adding an extra copy of the NR2B gene to the genome of a mouse called Doogy.[1] This choice of name was definitely not arbitrary, for just like the fictional television science prodigy Doogy Howser MD, the mouse was reported to acquire new knowledge at an unparalleled pace and retain it much longer than unmodified mice. Subsequent experiments with the NR2B gene yielded less convincing results, but the research team hopes that, once proved safe, this procedure could offer a revolutionary treatment for brain damage caused by Alzheimer's and Parkinson's. Even more spectacularly, they also speculate about what might be achieved by inserting additional copies of the NR2B gene in humans with normally functioning brains. Because if the treatment could restore damaged brains to health, then it could perhaps also be used to boost the capacities of the brain beyond its normal functioning.

In a recent article, the German behavioral geneticist Klaus-Peter Lesch claims to have identified a sequence of DNA on chromosome 5-HTTLPR that, in his view, has an important influence on our state of wellbeing.[2] His study shows that people who are born with a shorter version of this sequence are more susceptible to negative feelings and emotions such as depression, anxiety, and frustration. Those born with a longer version of the sequence, on the other hand, are reported to be more temperamentally upbeat and optimistic than the average person. Lesch still has no definitive explanation why this is the case, but there is good evidence that it has something to do with the fact that this sequence of DNA is responsible for regulating the reuptake of serotonin, the monoamine neurotransmitter also targeted by the antidepressant Prozac. But if his thesis proves to be correct, then it would be theoretically possible for parents to give their prospective children a guaranteed happy life by having only the longer version of this sequence engineered into their genome.

A number of recent announcements of progress and innovations in genetic science indicate that we are on the verge of entering a whole new era

1 Tang, Y., Shimzu, E., Dube, G. R., *et al.* (1999), 'Genetic enhancement of learning and memory in mice,' *Nature* 401: 63-69.

2 Lesch, K., Greenberg, B. D., Higley, J. D., *et al.*, 'Serotonin transporter, personality, and behavior,' in *Molecular Genetics and the Human Personality*, eds. J. Benjamin, R. Ebstein and R. Belmaker (Washington, DC: American Psychiatric Publishing, 2002).

in human evolution. Rapid advancements in genomics, most notably the completion of the Human Genome Project in 2000, have opened up the way for the manipulation of the human genetic heritage by technological means. In Jürgen Habermas's words, "what hitherto was 'given' as organic nature, and could at most be 'bred,' now shifts to the realm of artifacts and their production."[3] Although there are still a lot of practical barriers that have to be overcome before it will be possible to modify human DNA in the same way as DNA in plants and non-human animals can be modified, "the question is no longer whether we will manipulate embryos, but when, where, and how."[4] The most obvious application of human genetic technologies is in the prevention and healing of disease. It would, for example, become possible to insert healthy copies of a gene into cells that contain a defective gene so that the cells start to make the correct protein. However, as innovative as gene therapy may be, its sole aim is to restore diseased bodies to health, and as such it is still largely continuous with established medical practices. But this is far less clear, though, in the case of what is called "genetic enhancement," for here genetic knowledge about the genetic underpinnings of the phenotype is not used for treating disease, but to design human beings according to human will. It remains to be seen how radical these developments will prove to be, but, as biophysicist Gregory Stock maintains, it is already certain that "the enormous collective project of conscious human evolution has begun."[5]

While the technologies for genetic modification are quite recent, the attempt to enhance the biological characteristics of human beings is hardly unique to the "age of genomics," for clear parallels could be drawn with the ideals of eugenics.[6] Formulated in the late 19th century by Francis Galton, the term eugenics refers to both a branch of science and a popular movement that sought to improve human hereditary traits through a variety of social policies and practices. Although the content of the eugenic project varied considerably from country to country, it is best known for its support for policies of racial segregation and enforced sterilization in the United States and Nazi Germany. Although very little of the scientific basis on which these 20th century eugenic practices were premised withstands critical scrutiny, they do provide an important background to the contemporary politico-philosophical debate over the moral acceptability or otherwise of human genetic modification. The central

---

3 Jürgen Habermas, *The Future of Human Nature* (Cambridge: Polity Press, 2003), 12.

4 Gregory Stock, *Redesigning Humans: Choosing our Genes, Changing our Future* (Boston/New York: Mariner Books, 2003), 2.

5 *Ibid.*, 184.

6 See for example George J. Annas and Sherman Elias, *Gene Mapping: Using Law and Ethics as Guides* (Oxford: Oxford University Press, 1992).

assumption guiding much of the literature on the socio-ethical implications of new genetic technologies is that if they have eugenic effects, then they are also morally unacceptable. Recently, however, some commentators have taken a different approach to this issue, arguing that there is nothing intrinsically wrong with the goals of eugenics as such, and that its moral acceptability depends on the values and principles of the political ideology that regulates its implementation in society.[7] They reject as unjustified any comparison that might be drawn between the contemporary idea of human genetic modification and the Nazi eugenics program by indicating that the "new eugenics" will be firmly rooted in the liberal principles of individual autonomy and value plurality. Their central claim is that if the state remains neutral in this matter and does not intervene to enforce a particular conception of the good to be sought through genetic modification, then a "liberal eugenics" will greatly reinforce the freedoms associated with reproduction. Critics, on the other hand, claim that any attempt to integrate the eugenic ideal into a liberal framework is bound to fail and that it will corrupt the central tenets of liberalism to the point of turning it into its opposite.[8] Although this argument takes many different forms, the basic assumption is that human genetic enhancement should be rejected because it poses a potential threat to human nature. Moreover, since the idea of an immutable human nature is intrinsic to the notion of human dignity, this threat extends to the whole meaning of what it is to be human.

Although it is not our intention to intervene directly into this contemporary debate over human genetic enhancement by taking position, it nonetheless constitutes the starting point and focus of the present book.

---

7 See for example Agar, *Liberal Eugenics*; John Harris, *Enhancing Evolution: The Ethical Case for Making Better People* (Princeton: Princeton University Press, 2007); Lee M. Silver, *Remaking Eden: How Genetic Engineering and Cloning Will Transform the American Family* (New York: Avon Books, 1998); Ronald Dworkin, 'Playing God: Genes, clones and luck,' in *Sovereign Virtue: The Theory and Practice of Equality* (Cambridge, MA: Harvard University Press, 2000); Jonathan Glover, *What Sort of People Should There Be?* (New York: Penguin Books, 1984); Philip Kitcher, *The Lives to Come: The Genetic Revolution and Human Possibilities* (London: Penguin, 1996); Nick Bostrom (2003), 'Human genetic Enhancements: A Transhumanist Perspective,' *Journal of Value Inquiry* 37 (4): 493-506.

8 See for example Francis Fukuyama, *Our Posthuman Future. Consequences of the Biotechnological Revolution* (London: Profile Books, 2002); Habermas, *The Future of Human Nature*; Michael J. Sandel, *The Case Against Perfection. Ethics in the Age of Genetic Engineering* (Cambridge MA: The Belknap Press of Harvard University Press, 2008), George J. Annas, *American Bioethics. Crossing Human Rights and Health Law Boundaries* (Oxford: Oxford University Press, 2005) and Leon Kass (2003), 'Ageless Bodies, Happy Souls: Biotechnology and the Pursuit of Perfection,' *The New Atlantis* 1: 9-28.

The question of human genetic modification currently remains just that – a *question*. The ultimate threshold has still not yet been crossed. Should humanity go beyond this point of no return? Both advocates and opponents of human genetic modification seem to be narrowing the scope of the debate to this mode of questioning, without realizing that they thereby base their position on a similar set of assumptions about the nature of and relation between the concepts of "life" and "technology" that prevents them from reflecting on the question of human genetic modification in a sufficiently radical way. Regardless of whether the question is answered in the negative or in the positive, in both lines of reasoning the underlying assumption is that the question of human genetic modification essentially concerns the implications of an epochal confrontation between two radically heterogeneous forces: human life on the one side and technology on the other. In what follows, we will argue that this bespeaks a thoroughly inadequate conception of the human-technology relation, one that obfuscates more than it clarifies. Yet, the disturbing fact that many still uncritically rely on such an instrumental conception of technology – even at the very moment that its impotency stands exposed – indicates that it is very hard when not even impossible to extricate oneself completely from its hold. Be that as it may, we will take a completely different approach to this issue, and will seek to argue in what follows that the very fact that it is possible to modify the human genome retroactively reveals that there never was anything like a human nature in the first place. Is the human being's proper dwelling place the realm of *physis* or the realm of *tekhnē*? Such a question might at first sight seem strange or even nonsensical, but it surely can no longer be left in abeyance if we want to stand up to the demands of our age.

The first chapter (*Enhanced Life*) will critically discuss the current politico-philosophical debate concerning the permissibility of a liberal eugenics and argue that none of the dominant positions on human genetic enhancement is entirely satisfactory due to the limited, monadic conception of the human that is adopted in these models. It will be argued that the positions of both advocates and opponents of a liberal eugenics are inconsistent on a conceptual level as, ultimately, both end up violating the very central liberal principle of individual autonomy that they each nonetheless pretend to defend. In particular, it will be shown that while the argument against a new eugenics necessarily entails a preemptive dehumanization of any potential enhanced form of life, the argument for it threatens to reduce any non-enhanced form of life to the status of "wrongful life" or a life not worth living.

In the second chapter (*Bare Life*), it will be argued that the contradiction that any attempt to either improve or protect human nature produces a form of life devoid of intrinsic value can only be properly understood when it is transcribed in terms of what Michel Foucault called biopolitics. Foucault's thesis holds that modernity saw the gradual substitution of the sovereign model of power by a regime of biopower, in the sense that the deployment, strengthening and improvement of the biological dimension of human existence became the principal object and target of modern political strategies. Both Giorgio Agamben and Roberto Esposito have, however, recently pointed out that Foucault left us with a paradox which he was never able to resolve. For how is it possible that the rising to dominance of a form of power that precisely seeks to protect and strengthen life also generated a political regime – Nazism – whose eugenic practices aimed at the suppression of life? After discussing Agamben's and Esposito's responses to this question, it will be argued that the future emergence of a responsible eugenics is predicated on a critical understanding of the conceptions of life that they have developed: on the one hand, the human being exposed to the violence of a politics *over* life ("bare" or "sacred" life) and, on the other hand, the human being who will become the subject of a politics *of* life to come ("form-of-life").

The third chapter (*Enframed Life*) starts from the still widely neglected fact that Heidegger's writings contain several critical remarks on what he considered to be the ultimate aim of "technological enframing," namely the "the artificial breeding of human material."[9] It will be argued that, contrary to what has been generally assumed, Heidegger didn't refer to these developments as a mere illustration of his theory of technology but that genetic intervention figured as the negative paradigm on the basis of which he developed this theory in the first place. For Heidegger, the danger of biotechnology does not consist in concrete empirical threats posed by technical interventions in the human genome but in the threat that man will completely forget his ecstatic belonging to Being. The question will be raised, however, how Heidegger can substantiate the claim that genetic modification puts man's essence at risk, given his well-known criticism of attempts to ground man's essence in a biological dimension. It will be argued that this suggests that Heidegger's philosophy is not completely anti-naturalistic, as has been generally believed, but that he was only critical of the "zoological" conceptions of man of the metaphysical humanist tradition, because it is on the basis of these conceptions that man currently understands himself solely in biogenetic terms.

The fourth chapter (*Natal Life*) will use Hannah Arendt's brief but poignant remarks in the prologue to *The Human Condition* about the

---

9 Martin Heidegger, *Four Seminars*, trans. A. Mitchell and F. Raffoul (Bloomington: Indiana University Press, 2003), 106.

advent of a biotechnological revolution as a starting point for a renewed reflection on her concept of natality. Arendt calls natality, the fact that human beings enter the world through birth, the central category of political thought. But how can she assert that being born conditions one to act freely if she also seems to maintain that it is through labor, not action, that human beings deal with their biologically conditioned processes? Expanding on Arendt's largely neglected references to the work of the German anthropologist Arnold Gehlen, it will be argued that the concept of natality precisely undoes any strict division between *zoē* and *bios*, the conditioned and spontaneous, because it names the radical co-implication of biological birth and politico-linguistic birth. On the basis of this new understanding of natality, it will be argued that technological intervention in human reproduction will not irrevocably infringe on the human capacity to make radical new beginnings.

The fifth chapter (*Prosthetic Life*) will draw on Bernard Stiegler's theory of originary technicity to argue that human life is phylogenetically as well as ontogenetically radically bound up with technical prostheses and thus that, in a sense, the human is always already technologically "enhanced." According to Stiegler, the human can be defined as neither a biological state nor a philosophical condition (mind, consciousness, etc.) because it comes into being through technics. To substantiate this claim, we will focus on Stiegler's reading of the French paleontologist André Leroi-Gourhan who has argued that the stone flint and the cerebral cortex co-evolved in a process of reflexive development, and thus that the appearance of homo sapiens is in the most literal sense an effect of the pre-hominid's tool-use. This will allow us to conclude that biogenetic intervention is neither unprecedented, nor "anti-natural," but that it has made the process of humanization possible in the first place. One of the most important consequences of seeing technology as the intrinsic condition of humanity will be that simply being "for" or "against" genetic modification will become an impossible position to sustain.

Chapter One

Enhanced Life

Redesigning Life

At the beginning of the 1970s, less than two decades after James Watson and Francis Crick had first described the basic structure of the DNA molecule and famously claimed to have unraveled "the secret of life," microbiologists Kathleen Dana and Daniel Nathans made a lesser-known yet equally groundbreaking discovery. They showed that the restriction enzyme *endonuclease R*, found in *Haemophilus influenzae* bacteria, can be used to cut DNA at specifically targeted locations.[1] This discovery, for which Nathans – together with Hamilton Smith and Werner Arber – received the 1978 Nobel Prize in Physiology or Medicine, was so remarkable that it would come to be regarded as the first major breakthrough on the path to the deliberate modification of the genome of a living being. A decisive step was taken approximately two years later by Dana and Nathans' colleagues, Herbert Boyer and Stanley Cohen,[2] who became the first scientists to succeed in using Nathans' cutting technique to isolate a DNA molecule and attach it to a plasmid, a piece of extrachromosomal DNA found in *E. coli* bacteria, using DNA ligases. Proceeding in this way, they were able to create what is known as a recombinant DNA molecule. The special characteristic of such an artificial DNA molecule is that it can be transferred to any other living organism, where – if all goes well – it will be transcribed, translated, and, ultimately, caused to produce a particular protein. Boyer later founded the biotech company *Genentech* and used recombinant DNA technology to produce synthetic human insulin. The possibilities inherent to deliberately altering an organism's genome will obviously allow for many more promising medical applications in the near future, however.

Yet, there is little doubt that the domain which will be most radically affected by these developments in genetics is that of human reproduction.

1 Kathleen Dana and Daniel Nathans (1971), 'Specific cleavage of simian virus 40 DNA by restriction endonuclease of *Hemophilus influenzae*,' *Proc. Natl. Acad. Sci. USA* 68: 2913-2917.

2 Stanley N. Cohen, Annie C. Y. Chang, Herbert W. Boyer and Robert B. Helling (1973), 'Construction of Biologically Functional Bacterial Plasmids *In Vitro*,' *Proc. Natl. Acad. Sci. USA* 70: 3240-3244.

It is true that over the last few decades, this domain has already undergone a number of major changes as a result of the introduction of new technologies such as artificial insemination and in vitro fertilization (IVF). But there are good reasons to assume that genetic modification technologies will bring about even more radical transformations. After all, unlike assisted reproduction technologies, human genetic modification technologies are not intended to improve the chances of new human beings' coming into being, but to allow us to make decisions about *what kind of human beings* come into being. That is to say, these technologies will give us the power to endow human beings with specific genetic traits and characteristics. Although these technologies are still very much in their infancy, speculation about their future applications has encompassed significant improvements in the therapies currently available for known genetic disorders and engineered immunity to serious diseases such as cancer and HIV as well as the creation of entirely new biological functions, such as flight and the breeding of a superior race of transhuman beings.[3] Whether some, all, or none of these predictions are actually realized remains to be seen. Perhaps scientists are simply too optimistic about where the current revolution in molecular genetics is headed. Nicolas Agar, for example, reminds us that at the beginning of the 20th century, many believed we would be using nuclear-powered vacuum cleaners and growing ready-to-serve portions of chicken by the end of the century.[4] Some of the predictions surrounding genetic technologies may also turn out to be products of overactive imaginations, but it could be argued that this is no reason to postpone critical reflection until there is more certainty about what is realistically possible. After all, as has been shown on many occasions, there is always the risk that scientific and technological developments occur so unexpectedly that any thorough discussion about their possible impact is rendered impossible. As Hans Jonas pointed out in the 1970s with regard to genetic engineering, "[s]ince no less than the very nature and image of man are at issue, prudence becomes our first ethical duty, and hypothetical reasoning our first responsibility."[5]

Yet, if the main question is not if but when these technologies will be available, it would probably be no exaggeration to say that we are on the threshold of a post-Darwinian era in evolution. The central assumption of (Neo-)Darwinism holds that selection only occurs at the genetic

---

3 See for example Andy Miah (2008), 'Engineering Greater Resilience or Radical Transhuman Enhancement?' *Studies in Ethics, Law, and Technology* 2 (1): 1-18.

4 Nicolas Agar, *Liberal Eugenics: In Defence of Human Enhancement* (Malden, MA/Oxford: Blackwell, 2004), 23.

5 Hans Jonas, *Philosophical Essays. From Ancient Creed to Technological Man* (Chicago and London: The University of Chicago Press, 1974), 141.

level, in the sense that variations acquired at the somatic level are not passed from one generation to the next. This idea was first formalized in the 19th century by the German biologist August Weismann, who argued that an impermeable barrier exists between the germ cell line and the line of somatic cells. According to Weismann, evolution is a one-way street: germ cells produce somatic cells and are heritable, but somatic cells exert no influence on germ cells and are therefore evolutionary inert. What counts in evolution is the genotype, not the phenotype. In modern biology, a revised version of Weismann's germplasm theory was formulated by Francis Crick, who called it the Central Dogma of Molecular Biology.[6] The Central Dogma describes the normal flow of biological information in an organism and states that the flow proceeds from DNA to RNA and then to proteins, but not in the reverse order. As there is no back-flow of information from the body to the genome, the only changes that can occur in the genome are random mutations and rearrangements. In other words, any changes that may occur in the body of an individual organism and any knowledge or skills it might acquire during its lifetime are inevitably and irrevocably lost to the species as a whole when the individual organism dies. However, if it is true that Neo-Darwinism is constructed entirely upon the premise that "the genetic program does not take lessons from experience,"[7] then it seems that the emergence of genetic technologies will cause the human species to enter a new phase in evolution. After all, if it is possible to introduce changes in the genome deliberately, then it appears that the genetic program can indeed "take lessons" from experience. Ultimately, as these technologies come to be applied more widely, evolution will no longer occur through natural selection, but through artificial selection.

However, some argue that while it may be true that these technologies are completely new, the goal of deliberate selection is not in fact as revolutionary as it first appears. Clearly, deciding whom we want to have children with already provides us with some measure of control over the genetic characteristics of our children. Moreover, the assisted reproductive technologies that are currently available can also be used for this purpose. A paradigmatic example of an attempt to use assisted reproduction technology for the singular purpose of wielding greater influence over the future of one's prospective child is the famous Repository for Germinal Choice. Founded in 1978 by the American millionaire Robert K. Graham, the Repository offered access to sperm samples from men who had

---

6 Francis Crick (1970), 'The Central Dogma of Molecular Biology,' *Nature* 227: 561-563.

7 François Jacob, *The Logic of Life*, trans. Betty E. Stillmann (New York: Pantheon Books, 1974), 30.

excelled in their fields to women who hoped that inseminating themselves with the sperm cells of a gifted scientist or successful athlete would give their children a good chance of developing the same traits as their donor-father. However, although it has been reported that at least some of the Repository children came to be about as successful as their donors in later life, many attempts to select specific characteristics have failed because they were not based on scientific knowledge of the genetic foundations underpinning the traits and characteristics the researchers aimed to produce in a child. Unless such procedures are combined with a certain amount of genetic technology, it is highly unlikely that they will yield the desired result. For this reason, but possibly also because the prices were simply too high, the Repository failed to attract enough customers to be profitable and eventually went out of business in 1999.

Perhaps the most fine-tuned procedure currently available for selective reproduction is preimplantation genetic diagnosis (PGD).[8] The general idea behind PGD is that a number of embryos are first created using IVF and subsequently tested genetically. On the basis of these test results, the prospective parents can choose which of the embryos will be retained for implantation. Of course, this technique presupposes that sufficient information is available about the genetic basis of the condition the parents are interested in. Assuming this is the case, though, PGD is clearly a much more refined technique for selecting a particular type of child than paying a visit to Graham's Repository. Inevitably, however, there are also some serious disadvantages to PDG. Any procedure that involves the selection of embryos for implantation will also entail a decision about what will happen to the embryos that are not selected. These may be stored and implanted at a later date; used for research purposes; or, given that they possess certain undesired characteristics, simply discarded. This issue raises a number of legal and ethical questions which we shall not explore here.[9] It should be clear, though, that PGD requires that parents make a selection not among different versions of *one* possible future child, but among *different* possible children who already exist in a premature state. Moreover, PDG restricts prospective parents' choices to the types of possible future children that would normally be provided by nature; it does not allow them to design a child entirely according to their own wishes. If this were ever to be possible, human genetic modification technologies would be required.

8 See for example Anne McLaren, 'Prospects for pre-implantation diagnosis of genetic disease,' in B. Holland and C. Kyriacou (eds.), *Genetics and Society* (Wokingham: Addison-Wesley, 1993): 67-85.

9 See for example Günter Rager (2008), 'Is preimplantation genetic diagnosis ethically acceptable?' *Bioethica Forum* 1 (2): 81-88.

The Return of Eugenics

What Graham attempted to do with his Repository, what PDG – within certain limits – can accomplish with more precision, and what genetic modification promises to do much more effectively in the near future, is seize control of the genetic endowment of our descendants. It aims to change them for the better. Some political philosophers and bioethicists have argued that we should embrace this new control wholeheartedly and prepare for the advent of a new eugenic era. There is still much discussion among the advocates of a new eugenics regarding the precise meaning and scope of this term, but broadly speaking it can be said to promote the idea that prospective parents be given as much freedom as possible to decide which genetic traits their children will or will not possess upon birth. Such extensive reproductive freedom would allow parents not only to intervene in the genetic make-up of their future children for therapeutic purposes, but also to choose traits according to their personal preferences. Thus, in addition to the medical goals of preventing disease and restoring health, it would also allow for the enhancement of phenotypic characteristics such as intelligence, strength, height, and memory.

It is notable, however, that few advocates of a new eugenics are willing to call the practice they support by that name. John Harris, for example, prefers to speak of "deliberate selection"[10] and Gregory Stock favors the term "human self-design."[11] These authors' reluctance to use the term "eugenics" obviously has much to do with the shadow that still hangs over earlier attempts to make improvements to the biological foundations of human life. Many critics are indeed afraid that the emergence of a new eugenics will also prompt the return of some of the horrific acts committed in the field's name, such as the atrocities committed by Nazi eugenicists. This association is so strong that occasionally, when a practice is referred to as eugenic, it is in fact being described as morally reproachable.[12] The *Charter of Fundamental Rights of the European Union*, for example, outlines the "prohibition of eugenic practices, in particular those aiming at the selection of persons,"[13] without explaining what such practices are and why they would pose a threat to human dignity. Apparently, the mere use of the label is enough to indicate that the laws relate to a field of practices

10 Harris, *Enhancing Evolution*, 4.

11 Stock, *Redesigning Humans*, 3.

12 See for example Stephen Wilkinson (2008), '"Eugenics talk" and the language of bioethics,' *Journal of Medical Ethics* 34: 467-471; Diane B. Paul (1992), 'Eugenic Anxieties, Social Realities, and Political Choices,' *Social Research* 59 (3): 663-683.

13 *Charter of Fundamental Rights of the European Union* (2000/C 364/1), Chapter 1, Article 3 (Right to the integrity of the person).

that any reasonable person would find morally objectionable. Conversely, those who argue in favor of a new eugenics are almost invariably accused of offering a thinly veiled justification of Nazism. Clearly, much of this argument is largely rhetorical. Even the most radical opponents of a new eugenics are aware that eugenics cannot be reduced wholly to its Nazi variant. Nevertheless, even a brief look at the history of eugenics reveals that there are good reasons to be wary of the call for a new eugenics.[14]

The term eugenics (from the Greek "good birth") first emerged at the end of the 19th century in the work of Francis Galton – a cousin of Charles Darwin – who defined it as follows:

> The science of improving stock, which is by no means confined to questions of judicious mating, but which, especially in the case of man, takes cognizance of all influences that tend in however remote a degree to give the more suitable races or strains of blood a better chance of prevailing speedily over the less suitable than they otherwise would have.[15]

Galton's original intuition was that the unequal distribution of intelligence among family pedigrees could be entirely explained by heritable factors. He therefore proposed that it should be possible to improve the overall level of intelligence in a population by encouraging members of favored families to have more children and discouraging members of less favored families from doing so. Galton's influence was almost immediate and within the next few decades, eugenics societies were established all across Europe and the Americas. Eugenics was both a respectable academic discipline and a popular movement. While many universities were beginning to offer courses on eugenic science and collect data on family pedigrees, eugenicists were also organizing popular fairs and exhibitions and even holding "fitter families" and "best baby" competitions. Galton himself was only interested in the hereditary factors that might influence intellectual virtue, but an enormous number of similar studies carried out in his wake subsequently claimed to have found the hereditary basis of many other characteristics and behaviors. Thus, as the movement grew and also began to attract attention in political circles, eugenics swiftly

---

[14] For an excellent and extensive historical account of the eugenics movement, see Daniel J. Kevles, *In the Name of Eugenics. Genetics and the Uses of Human Heredity* (Cambridge MA: Harvard University Press, 1985). The following overview largely draws on the useful summary contained in Allen Buchanan, Dan W. Brock, Norman Daniels and Daniel Wikler, *From Chance to Choice. Genetics & Justice* (Cambridge: Cambridge University Press, 2000), 30-40.

[15] Francis Galton, *Inquiries Into Human Faculty and its Development* (London: J. M. Dent & Company, 1883), 17.

became a catch-all term that could refer to all kinds of practices and measures for improving the biological characteristics of a population.

Yet, despite the wide variety of meanings attributed to the term, it could be said that all eugenicists accepted three basic assumptions. First, they believed that populations were being threatened by a growing process of degeneration. It was thought that welfare programs, improved medicine, and better living conditions in general gave those who would normally be on the losing side of the Darwinian fight for survival an "unnaturally" high chance of persevering in existence. In addition, as these "deficient" persons tended to reproduce in larger numbers, the net result for the population would be a steady decline of those endowed with "favorable" characteristics. Second, it was assumed that behavioral characteristics were heritable. This was the case not only in France, where most eugenicists still accepted the Lamarckian theory of the heritability of acquired characteristics, but also in countries where eugenicists adhered to a more standard version of Darwinism. Finally, eugenicists also agreed on the overall aim of eugenics. They believed that a better understanding of the mechanisms of human heredity would enable scientists and politicians to take measures towards improving the overall quality of the population's gene pool.

There was less agreement amongst eugenicists, however, about how this goal was to be achieved. Galton himself would not go further than the idea of encouraging the fittest to have large families but there were many other eugenicists who believed that more drastic measures were necessary to put a stop to the creeping process of degeneration. In the United States and elsewhere, laws were being passed to legalize the involuntary sterilization of the "feeble-minded." In the famous *Buck v. Bell* case, for example, Supreme Court Justice Oliver Wendell Holmes rejected an appeal made by 18-year-old psychiatric patient Carrie Buck who claimed that, by ordering her sterilization, the state of Virginia had violated her constitutional rights. In his ruling, Holmes defended his decision by arguing that:

> It is better for all the world if, instead of waiting to execute degenerate offspring for crime or to let them starve for their imbecility, society can prevent those who are manifestly unfit from continuing their kind. The principle that sustains compulsory vaccination is broad enough to cover cutting the Fallopian tubes. [...] Three generations of imbeciles are enough.[16]

In the light of such cases, it is unsurprising that Nazi eugenicists were quick to point out that the United States served as a model for their

[16] *Buck v. Bell*, 274 U.S. 200 (1927).

own sterilization program. Of course, Nazi Germany's program of racial hygiene surpassed all other eugenic policies in both scale and cruelty; no other regime has ever concluded that the only means of achieving its eugenic goals was to eliminate the "unfit elements" of its population. However, it is certainly not the case that authoritarian regimes are the only ones to have supported coercive measures. As mentioned above, similar sterilization laws were adopted in the United States but also in countries where the social democrats were in power, such as Sweden and Denmark. Moreover, the 1935 Nuremberg Laws on Citizenship and Race were not the first government regulations enacted on the basis of eugenic concerns. By the beginning of the 20th century, most states of the USA had already adopted miscegenation laws. Later, following World War One, the United States would drastically restrict immigration from Southern and Eastern European countries out of the fear that the mass arrival of newcomers with inferior genes would negatively affect the American gene pool. Furthermore, much of the research carried out on "bad genes" relied on poor science and were often nothing more than simple reflections of race and class biases. Deficient traits and behaviors which were thought to have a hereditary basis included not only immorality, criminality and even poverty, but also character and talent. It is unlikely that more thorough scientific research into the hereditary foundations of certain traits and behaviors would have made a decisive difference, since the science of genetics was still very much in its infancy. Moreover, many of the leading academics in the eugenics movement were actually social scientists, who were looking for statistical correlations between social position and what they perceived to be inherited capabilities.

In view of these disconcerting historical facts, it is perhaps unsurprising that eugenics is so often depicted in negative terms. Yet despite its inglorious history, the concept of eugenics continues to attract enthusiastic supporters. There appears to be something undeniably appealing in the essential idea of eugenics, something that prevents us from rejecting it in its entirety. Who, after all, would not want to give his or her child the best possible genetic endowment? Convinced that the potential benefits of genetic technologies in human reproduction are too valuable to renounce on the basis of past abuses, advocates of a new eugenics argue that the main question is not whether the Nazi eugenics program was abhorrent – a question to which no sane person could respond in the negative – but whether the atrocities committed in the name of eugenics were not in fact the result of the underlying Nazi ideology rather than something intrinsic to the field of eugenics itself. Provided that the eugenic goal of improving human beings still enjoys universal support and approval, and that the moral acceptability of eugenics depends on the values and principles of the political ideology regulating its implementation in society, then, they suggest, it might still be possible to devise a form of eugenics that is compatible with the fundamental tenets of contemporary liberal democracy.

The Liberal Eugenics

Nicholas Agar, one of today's most vocal advocates of a new eugenics, has recently argued that the central principles of liberalism provide ample guidance for avoiding the moral pitfalls of earlier forms of eugenics: "[T]he addition of the word 'liberal' to 'eugenics' transforms an evil doctrine into a morally acceptable one."[17] In his view, the most important difference between the authoritarian eugenics of the past and the liberal eugenics he envisages is simply the degree of control that the state has over the reproductive choices of its citizens: "While old fashioned authoritarian eugenicists sought to produce citizens out of a centrally designed mould, the distinguishing mark of the new liberal eugenics is state neutrality."[18] Robert Nozick was perhaps the first liberal philosopher to argue for a decentralized eugenic policy. In *Anarchy, State and Utopia* he suggests that in the coming age of genetic engineering, reproduction could best be regulated by creating a genetic supermarket: "This supermarket system has the great virtue that it involves no centralized decision fixing the future human type(s)."[19] State neutrality is, of course, central to any liberal democratic system which aims to protect the principle of value pluralism. Liberal democratic societies are founded on the beliefs that, firstly, individuals may have radically different ideas about what makes a good life, and secondly, a government should not interfere with the choices made by its citizens in pursuing these ideas. In its original formulation, the principle of value pluralism was primarily intended to safeguard freedom of religion and expression, but liberal eugenicists believe that it is broad enough to cover the freedom to use genetic technologies in the field of reproduction.[20] This means that governments must refrain from interfering not only with the more ordinary reproductive choices of its citizens – as whom they want to have children with, how many children they want to have, and indeed whether they want to have children at all – but also with new reproductive choices made possible by genetic technologies – such as what kind of children they want to have. In its negative form, this principle ought to prevent a government imposing specific reproductive goals on its citizens, including the shaping of future

17 Agar, *Liberal Eugenics*, 135.

18 Nicholas Agar (1998), 'Liberal eugenics,' *Public Affairs Quarterly* 12 (2): 137.

19 Robert Nozick, *Anarchy, State and Utopia* (New York: Basic Books, 1974), 315.

20 See John A. Robertson, *Children of Choice: Freedom and the New Reproductive Technologies* (Princeton: Princeton University Press, 1994).

human beings according to an "ideal type." In its positive form, it gives prospective parents the freedom to consult their own values in deciding which genetic traits they want to pass on to their children. As Agar sums it up, "[a]uthoritarian eugenicists would do away with ordinary procreative freedoms. Liberals instead propose radical extensions of them."[21]

One of the most important reasons why liberal eugenicists are convinced that there is no need for moral panic in the face of a new eugenic era is that genetic interventions are not radically different from other types of interventions made in human beings. That is to say, they see no morally relevant difference between shaping humans by making modifications to their environment and shaping humans by making modifications to their genes. There are two aspects to this claim. First, it allows liberal eugenicists to refute the common argument that genetic intervention is substantially more intrusive than any other influence we may have over the development of another human being. Agar calls this the nurture principle: "If we are permitted to produce certain traits by modifying our children's environments, then we are also permitted to produce them by modifying their genomes."[22] Note that this principle does not serve as a justification for unlimited genetic modification but only suggests that such modifications do not pose radically different problems from those pertaining to other influences parents already have over their children's development. Consider the example of improved intelligence. If parents are allowed and even encouraged to increase their children's intelligence by providing them with the best possible education, why, then, should they not be allowed to pursue the same goal by means of genetic technologies? Second, if there is no substantial difference between genetic intervention and other influences that parents have over the development of their children, then there is also no need to develop new ethical guidelines and legal regulations for genetic technologies, because the freedom to use such technologies is already protected by the existing right to reproductive freedom. Moreover, as John Harris has argued, this right does not merely involve "the exercise of bare preference, like drinking coffee or playing tennis," but could be compared to "a fundamental human right."[23] Positioning reproductive freedom alongside other fundamental rights, such as freedom of religion, Harris concludes that parents should enjoy a maximum of freedom in the domain of reproduction and thus that any arguments proposed against genetic enhancement would need to be of a very high order to overrule this right.

---

21 Agar, *Liberal Eugenics*, 137.

22 *Ibid.*, 113.

23 Harris, *Enhancing Evolution*, 75.

Liberal eugenicists also argue that the nurture principle is supported by genetic research, in the sense that it has been convincingly shown that any strong notion of genetic determinism is unfounded. Fears about genetic modification having the power to fully dictate a person's characteristics and behaviors largely spring from the mistaken view that genes are exclusively responsible for the shaping of a person. Critics of human genetic modification sometimes presume that, for example, simply inserting an extra gene for athletic ability will produce an individual fated to become a successful athlete in later life. Leaving aside for a moment the question of whether such a gene really exists, it is important to note that this view of genes' influence on a person's development is not supported by scientific evidence. Cases of monozygotic twins have amply shown that two human beings with identical genomes do not grow to become identical persons. The explanation for any physical and psychological differences between them must therefore be sought in the environmental conditions to which they have been exposed. This implies that neither genetic factors nor environmental factors are entirely responsible for a person's traits, but that these result from complex interaction between the two. Thus, unless a person endowed with a gene for athletic ability is also encouraged to engage in sporting activities, given access to adequate sporting facilities and subjected to a strict training schedule, she will not have significantly higher chances of becoming a successful athlete than a random person who has not been genetically engineered for athletic ability.

However, even if we provisionally grant that genetic modification is no more morally problematic than any other kind of attempt to influence the development of a human being, does this mean that there should be no specific limitations whatsoever in the use of genetic technologies? Would it not be morally wrong, for example, to use these technologies to deliberately produce a child with a severe physical or psychological disorder? If we are spontaneously inclined to say that this would indeed be wrong, we must keep in mind that some would argue that even deliberately selecting for a disability is no straightforward case of an immoral act. Consider the example of the American deaf lesbian couple Candace McCullough and Sharon Duchesneau, who hoped to get pregnant using the sperm of a deaf donor in order to increase the chances that their child would also be born deaf.[24] After having had their request rejected by several sperm banks, they approached a deaf friend who agreed to donate his sperm. Following several unsuccessful attempts, Duchesneau gave birth to a daughter who is reported to have slight hearing in one ear. Would a liberal eugenics

---

24 David Teather, 'Lesbian couple have deaf baby by choice,' *The Guardian*, 8 April 2002: 2.

allow this couple to use genetic modification technologies to deliberately create a deaf child? It seems that even a liberal eugenicist could only answer this question in the negative, because the parents would apparently be causing harm to their child. According to Agar, however, it is not quite that simple. Suppose that Duchesneau was not a lesbian and that she was in a relationship with the deaf sperm donor. If they decided to have a child, then there would be an equally high chance that this child would be born deaf. If that were to happen, Agar asks, would we then also say that this couple has intentionally harmed their child?[25] There is another issue to be considered here, however. The deaf couple themselves are not of the opinion that they have created a disabled child. They argue that, from their point of view, hearing people can also be considered disabled because they are unable to participate fully in the rich culture of the deaf. In other words, where some would consider this to be a straightforward case of a harmed or disabled condition, others would view it as a divergent yet equally valuable way of life. Perhaps the liberal commitment to state neutrality should also apply to this case.

It has often been argued that some of these borderline issues could be avoided by drawing a line between therapy and enhancement.[26] There are two assumptions inherent to this argument. The first is that there is an objective difference between genetic interventions that aim at restoring the capacities of the body to their "normal" state and interventions that aim at raising them above this state. The second assumption is that this distinction corresponds to the moral boundary between permissible and impermissible uses of genetic technologies. In other words, this argument holds that there is nothing morally wrong with using genetic technologies to heal people, but that it is impermissible to use them to boost human capacities above what is normal, or for that matter, below what is normal, as in the case of deafness mentioned above.[27] While this argument appears to possesses the merits of simplicity and fitness for practical application, both of its assumptions have met with severe criticism from liberal eugenicists. First, while the difference between a diseased state and non-diseased state enjoys great intuitive plausibility, this distinction may prove harder to sustain conceptually. At the risk of gross simplification, it could be said that there are two approaches to the definition

[25] Agar, *Liberal Eugenics*, 13-14.

[26] See for example LeRoy Walters and Julie Gage Palmer, *The ethics of Human Gene Therapy* (New York: Oxford University Press, 1997), xviii and Alastair Campbell, Grant Gillett and David Gareth Jones, *Medical Ethics*, 2nd edition (Oxford: Oxford University Press, 1999), 76.

[27] Jackie Leach Scully (2001), 'Drawing a Line: Situating Moral Boundaries in Genetic Medicine,' *Bioethics* 15 (3): 189-204.

of disease. There is the social constructivist approach, which essentially defines diseases as states which the majority of society considers harmful. However, it is clear that decisions about what constitutes a disease will in some cases simply reflect social prejudice. The objectivist approach, on the other hand, defines disease as any departure from normal biological function. This definition is advantageous in that it makes popular attitudes irrelevant but it could also result in the banning of a number of applications of genetic technology which nonetheless promise to yield great health benefits. For example, if it were possible to genetically engineer immunity to cancer, then a ban on using genetic technologies to enhance humans beyond their normal state would also include this kind of genetic intervention.

Others have argued that while it may be true that the boundary between therapy and enhancement is too unstable to serve as an absolute criterion for moral judgment, there are still good reasons not to abandon it altogether. Buchanan, Brock, Daniels and Wikler agree that this distinction is indeed not entirely satisfactory, epistemologically speaking, but they suggest that any commitment to a Rawlsian liberal notion of justice as fairness requires that we nonetheless retain some limited version of it. Their basic assumption is that if a just liberal society rightly seeks to establish equal opportunities by compensating those whose initial, disadvantaged position in society is due to social factors over which they have no control, then it should do likewise for those whose opportunities are limited by genetics. In other words, according to these authors, the liberal principle of fairness should be applied not only to inequalities of opportunity that result from the "social lottery," but also to inequalities that result from the "natural lottery:" "As our powers increase, the territory of the natural is annexed to the social realm, and the new-won territory is colonized by ideas of justice."[28] Specifically, this would mean that a just society is obliged to provide genetic technologies to those suffering from a genetically induced condition that will prevent them from participating fully in society. Following this argument, the therapy/enhancement distinction will not correspond entirely to the permissible/impermissible boundary, but given resource constraints, it does serve as a key rationale for the fair distribution of the benefits of genetic technologies.

Central to the model devised by Buchanan and his co-authors is Christopher Boorse's notion of "normal species functioning." Boorse claims to be able to distinguish objectively between a state of health and a state of disease by defining the former as "statistical normality of function, i.e., the ability to perform all typical physiological functions with at

[28] Buchanan *et al.*, *From Chance to Choice*, 84.

least typical efficiency."[29] Normal functioning is thus understood as the ability to execute biological functions which, from a statistical point of view, are typical of a particular species. Accordingly, diseases are "internal states that depress a functional ability below species-typical levels."[30] Since the notion of normal functioning merely refers to a statistically ideal state, every individual will deviate from it to a certain degree. Not all deviations from the norm will necessarily be defined as diseases: only those that fall outside of the range of what is statistically considered normal will count. However, Buchanan and his co-authors are not primarily interested in whether the notion of normal species functioning allows for a non-normative definition of disease. Instead, they assert that it provides us with a fair criterion for deciding whether or not a particular genetic condition requires technological intervention. After all, they explain, what matters "from the standpoint of a general account of equal opportunity, is not whether they are diseases but whether they limit opportunity."[31] That is to say, a just society should not restrict the application of genetic technologies to the treatment of disease, but allow it for all departures from normal functioning that deprive people of "the ability to participate in political, social, and economic life."[32] The normal function model of health care that they propose is therefore both more restrictive and more extensive than models that rely on the treatment/enhancement distinction alone: more restrictive because it forbids interventions in departures from normal functioning that do not impact on participation, and more extensive because it allows for interventions in genetic conditions that fall within the range of normal functioning if these conditions prevent the individual from participating "as normal competitors in all spheres of social life."[33]

John Harris, one of the most fervent advocates of a new eugenics, has fiercely criticized the appeal to equality of opportunity in justifying the use of genetic technologies for enhancement purposes. Surely, since we would also wish to use these technologies to address conditions which do not affect participation, the moral reasons we have for using them are not that they reestablish equality of opportunity – although this may constitute an additional reason – but that they change us for the better. Moreover, in his view, "enhancements are not plausibly defined relative to normalcy, to

[29] Christopher Boorse (1977), 'Health as a Theoretical Concept,' *Philosophy of Science* 44 (4): 543.

[30] *Ibid.*

[31] Buchanan *et al.*, *From Chance to Choice*, 72.

[32] *Ibid.*, 122.

[33] *Ibid.*

normal species functioning, nor to species-typical functioning."[34] These notions "play no part in the definition of harm and therefore no part in the way the distinction between therapy and enhancement is drawn."[35] He gives a striking example to illustrate this. Suppose it was possible to use genetic technologies to slow down or even halt the ageing process. If we followed Boorse's conception of normal functioning, then we would have to forsake this clearly benevolent use of genetic technology because it would not simply restore our body to normal functioning but actually enhance it beyond its normal state. In other words, since it is perfectly normal for us to die of the diseases of old age, this intervention would go beyond the therapeutic use of genetic technologies and would therefore be morally unacceptable. As a libertarian consequentialist, Harris believes that it would be better to do away with the therapy/enhancement boundary altogether. In his view, the moral imperatives either to provide therapy or enhancement derive from the fact that we value minimizing harm and maximizing benefits. What counts in deciding if it would be permissible to use genetic technologies is not the fact of whether an individual's current state deviates from normal functioning, but the cost/benefit calculation regarding the body's "possible functioning."[36] That is to say, the only pertinent questions are whether the harm the technologies aim to prevent is serious enough and whether the benefits they aim to produce are valuable enough to take the risks.

Life needs to be Protected

Some critical commentators have argued, however, that the proposed marriage between eugenics and liberalism will not so much redeem the former of its authoritarian drift as corrupt the central principles of the latter to the point of its becoming something different altogether. According to these critics, genetic enhancement is morally wrong not only because it is still unsafe or because it goes beyond the medical goal of restoring the body to its normal biological functioning, but also, and more grandly, because it threatens to change human nature.[37] This argument can take a variety of different forms, but the basic assumption is that modifications to the human genome threaten to disrupt something that is valuable in itself. Consequently, since our very understanding of human dignity and its legal reflection in human rights is founded upon the notion of human

34 Harris, *Enhancing Evolution*, 36.

35 *Ibid.*, 46.

36 *Ibid.*, 53.

37 See Kass, *Ageless Bodies*, 14-18.

nature, then genetic modification could ultimately signal the end of some of the central principles of liberal democracy.

George Annas is perhaps the most outspoken adherent of this approach. Arguing that genetic modification will create a species of creatures which we will no longer be able to recognize as fellow human beings, he predicts that "either the normal human beings will view the 'better' humans as the other and seek to control or destroy them, or vice-versa."[38] This prospect makes the project of a new eugenics seem not merely morally problematic, but actually repulsive: "It is this genocidal potential that makes some species-altering genetic engineering projects potential species-endangering weapons of mass destruction and the unaccountable genetic engineer a potential bioterrorist."[39] Eugenic programs working in this direction could, as a consequence, "fit into a new category of 'crimes against humanity' in the strict sense, actions that threaten the integrity of the human species itself."[40] While Annas is well aware that we are still a long way from realizing such sweeping genetic interventions, he argues that we should already be deciding upon the measures necessary for preventing any further steps being taken toward that goal, in the form of an international, legally binding "human species protection treaty."[41]

Yet, it is one thing to say that we are on the verge of losing something essential to being human and quite another to specify what it is that that essence consists of, precisely. Throughout the ages, philosophy and theology have provided numerous definitions of human nature by identifying it with soul, ratio, language, moral freedom, or some other distinctive quality. The question is, however, whether the notions traditionally used to delineate human beings from other living beings are still meaningful in the present context. Take, for example, the Christian belief that human beings enjoy a superior status among the Earth's creatures because they alone were created in the image of God and endowed with an immortal soul. Clearly, this idea of human nature is central to many Christian thinkers' conservative reaction to a number of contentious bioethical issues such as cloning, stem-cell research and genetic modification. This is precisely where things become particularly confusing, though, since if the soul is truly the essence of the human being, then why should we be afraid that interventions in the mortal body will wreak havoc on human nature? In Slavoj Žižek's words, "if the church effectively believes in the immortality of the human soul, in the uniqueness of the human

38 Annas, *American Bioethics*, 38.

39 *Ibid.*

40 *Ibid.*, 40.

41 *Ibid.*, 41.

personality, in how I am not just the result of the interaction between my genetic code and my environs, then why oppose cloning and genetic manipulations?"[42] Arguably, this question can also be posed to Kantian-inspired thinkers who are critical of genetic modification on the grounds that it would amount to treating human beings merely as means and not as ends in themselves. If it is true that human beings possess dignity because only they have the ability to transcend the causal realm of nature and act according to moral law, then why oppose attempts to bring the natural foundations of human life under the control of the free subject?

Francis Fukuyama has argued, however, that while science has made it abundantly clear that what were previously thought to be unique, markedly human qualities – such as language and reason – are in fact merely the contingent results of long evolutionary processes and also that these are, at least to some degree, also present in other living beings, there are nonetheless convincing reasons not to reject the notion of human nature altogether. The most important of these is that the belief that there is something that unites all human beings is fundamental to the central liberal principle of universal human equality. This principle states that while individual persons may differ from each other in many different ways, in their languages, professions, talents, religions, degrees of wealth, genders and so on, there is something that nevertheless makes them equal and that is more important than these individual characteristics. This "something" is what Fukuyama calls "Factor X," the most basic element of what it means to be human. All human beings are equal because each of them possesses Factor X. The use of new genetic technologies could change this, however. In liberal democracies we tend to accept some of the social inequalities that might result from the uneven distribution of genetic dispositions because we know that initially, we all had an equal chance of "winning" or "losing" the genetic lottery. In fact, one of the reasons why social redistribution mechanisms have been created is that we acknowledge that social success is partly due to influences beyond the control of the individual. Moreover, the genetic lottery also has a redistributing function of its own: children of successful individuals will not necessarily inherit the gene-based talents of their parents, and the children of underprivileged individuals may inherit talents that will allow them to climb further up the social ladder. But, Fukuyama predicts, "when the lottery is replaced by choice, we open up a new avenue along which human beings can compete, one that threatens to increase the disparity between the

42 Slavoj Žižek, 'Of Cells and Selves,' in *The Žižek reader*, eds. Elizabeth Wright and Edmond Wright (Malden Mass.: Blackwell, 2004), 306.

top and bottom of the social hierarchy."[43] When the socially privileged no longer pass on just social advantages to their children but also endow them with genetic dispositions that give them an unfairly high chance of securing an advantageous position in the social competition, these enhanced individuals "may come in time to think of themselves as different kind of creatures" and "feel themselves to be aristocrats, and unlike aristocrats of old, their claim to better birth will be rooted in nature and not convention."[44]

What, then, is this Factor X which all human beings are supposed to have in common and which entitles them to equal dignity and rights? Fukuyama agrees that our claim to equality can no longer be grounded in a transcendent understanding of Factor X, but neither does he think that we can ground universal human equality in positive law, for it is unlikely ever to be possible to reach cross-cultural agreement on the definition of Factor X. Thus, in his view, "there is desperate need for philosophy to return to the pre-Kantian tradition that grounds rights and morality in nature."[45] Accordingly, Fukuyama defines human nature as "the sum of the behavior and characteristics that are typical of the human species, arising from genetic rather than environmental factors."[46] Notice that this definition does not require all individuals to have an equal share in these characteristics, nor does it exclude the possibility that some will lack one or more of them without necessarily being denied the status of human being. This is so because, firstly, typicality is a statistical concept that refers to a median of a normal distribution of characteristics and traits (with its limits set by nature) and, secondly, because this definition also applies to human beings who might lack one or more of these characteristics, but have at least the theoretical potential to have them (e.g. children, the disabled, etc.). According to Fukuyama, every human being is born with a set of genes that allows him or her to behave in a species-typical way, in the sense that all humans are biologically equipped to "see colors, react to smells, recognize facial expressions, parse language for evidence of deceit, avoid certain dangers, engage in reciprocity, pursue revenge, feel embarrassment, care for our children and parents,"[47] and so on. It is on the basis of these genetically programmed traits and behaviors that human beings are able to develop the higher qualities that are usually seen as distinctively human, such as reason, moral freedom and sociability.

43 Fukuyama, *Our Posthuman Future*, 157.

44 *Ibid.*

45 *Ibid.*, 112.

46 *Ibid.*, 130.

47 *Ibid.*, 143.

However, Factor X cannot be reduced to any one of these individual qualities: "It is all of these qualities coming together in a human whole that makes up Factor X."[48] This allows Fukuyama to state more clearly what is wrong with human genetic modification, because to the extent that genetic technologies will allow us to modify one or more of these traits and characteristics beyond the level of species-typicality, they will also threaten to disrupt the precarious balance among the qualities that underpin Factor X. Thus, Fukuyama concludes, if liberal democracy wishes to remain committed to a strong notion of universal human equality, it is urgently necessary that we create a new international regulatory organ that "not only would have to have a mandate to regulate biotechnology on grounds broader than efficacy and safety but would also have to have statutory authority over all research and development."[49]

In his book-length essay *The Future of Human Nature*, Jürgen Habermas argues in a similar vein that human nature should be legally protected against genetic enhancement. Yet he founds this claim on a very different understanding of what it essentially means to be human. As liberal societies foster the belief that science and technology promote individual freedom and autonomy, Habermas is under no illusion that a return to a metaphysical or religious understanding of human nature as somehow sacred in itself might curtail the further development of technologies which seem to offer the prospect of increased health, prolonged lifespan and other benefits that will enlarge the scope of personal choice. Neither does he believe, however, that precise moral boundaries can be drawn on the basis of medico-scientific arguments about the dividing line between the scope of therapy and the scope of enhancement. For just as in the heated discussions surrounding the "beginning of life" and their relevance to issues having to do with interventions on prenatal forms of life, it will no doubt prove impossible to find unambiguous criteria which are sufficiently neutral with respect to the various "worldviews" and scientific paradigms circulating in the public sphere of liberal democracies. Habermas therefore agrees with Fukuyama that we are in desperate need of a post-metaphysical conception of human nature that would place human dignity on more solid ground. Unlike Fukuyama, though, what he seeks to protect are not the "species-typical" characteristics and behaviors of *homo sapiens*, but "the *conditions* under which the practical self-understanding of modernity may be preserved."[50] He argues that once we achieve a reflexive understanding of the necessary conditions for "our

48 *Ibid.*, 171.

49 *Ibid.*, 215.

50 *Ibid.*, 26.

capacity to see ourselves as the authors of our own life histories,"[51] we will realize that a liberal eugenics contradicts these conditions and should therefore be rejected.

Central to Habermas's argument is the notion that dignity is not a property one possesses simply by virtue of being human, but that it is the distinctive mode of being of a "communicatively structured form of life."[52] What he means by this is that we are only able to understand ourselves as free and autonomous agents worthy of respect in the context of a moral community that consists of equal members interacting with each other on the basis of norms and reasoning. This communal aspect of human nature is much more fundamental to our conception of human dignity than any of the more substantive elements that have been attributed to it. After all, it is only by virtue of the existence of a community of equals capable of engaging "in interpersonal relations of mutual respect"[53] that such normative claims can be defended at all. Thus, when Habermas states that the danger of genetic technologies lies in their power to change human nature, he means that their free deployment threatens to undermine the very foundations of the moral community.

To demonstrate why this is so, Habermas invites us to consider that our lifeworld is still largely "Aristotelian," in the sense that we tend to make automatic distinctions between "what is *manufactured* and what has *come to be by nature*,"[54] between the "artificially produced" and the "organically grown." This distinction is morally relevant insofar as it motivates us to adopt a particular mode of action when dealing with entities belonging to either one of these realms: while inert, inorganic entities are open to various forms of technical-instrumental intervention, self-regulated organic entities are not. According to Habermas, this is due to the fact that we spontaneously feel "empathy" for organisms which seem to possess a certain amount of subjectivity, no matter how minimal. We remain committed to this logic in the case of genetic interventions carried out on embryos for therapeutic purposes, firstly because our actions in this case are still guided by the natural processes of growth inherent to this prenatal form of life, but also because we imagine how the future person might give consent for any intervention that could prevent or cure a debilitating condition. In the case of genetic enhancement, however, a very different scenario emerges. Here, prospective parents are not treating the embryo as another subject who will come to be on an equal footing with them, but as an object they can simply

51 *Ibid.*, 25.

52 *Ibid.*, 72.

53 *Ibid.*, 33.

54 *Ibid.*, 46.

dispose of if necessary. In other words, if the relationship between parent and child is reduced to that of producer and product, they will never be able to meet each other as equal members of the moral community.

According to Habermas, then, liberal eugenicists make the mistake of focusing solely on the freedom enacted in parental choice, while the proper question to ask is what consequences genetic intervention will have for the programmed person's "capacity of being oneself"[55] on which one's ethical self-understanding as a free and autonomous member of a liberal egalitarian society depends. It is true that if "we experience our freedom with reference to something which, by its very nature, is not at our disposal,"[56] then the situation of the programmed person is not fundamentally different from that of an individual born the "natural" way, for neither have had any say in the genetic traits and characteristics they are endowed with. The crucial question to ask, however, is if it makes any difference whether these traits are the result of natural chance or of the deliberate intervention of a third person. Liberal eugenicists tend to play down the impact of this intervention on the existential situation of an enhanced individual by suggesting that there is no substantial difference between improving a person by modifying her social environment and doing so by modifying her genes. For Habermas, however, this is exactly what makes all the difference. In his view, while a genetically unenhanced person always retains the option of rejecting or reappraising her parents' attempts to shape her personality through socialization, the enhanced person "who is at odds with genetically fixed intentions is barred from developing [...] an attitude towards her talents (and handicaps) which implies a revised self-understanding and allows for a productive response to the initial situation."[57] Moreover, a liberal eugenics would not only deprive the genetically enhanced person of the spontaneous self-perception of being the singular author of her own life, but also create the child's permanent and irreversible social dependence on the parent, which "is foreign to reciprocal and symmetrical relations of mutual recognition proper to a moral and legal community of free and equal persons."[58] As Habermas has already argued with regard to human cloning, such an asymmetrical relationship would surpass even slavery in cruelty, for it would never be possible to emancipate oneself from this inferior, programmed position.[59]

55 *Ibid.*, 57.

56 *Ibid.*, 58.

57 *Ibid.*, 62.

58 *Ibid.*, 65.

59 See Jürgen Habermas, 'An Argument against Human Cloning. Three Replies,' in *The Postnational Constellation. Political Essays*, trans. and ed. M. Pensky (Malden, Mass.: Polity Press, 2007), 163-172.

The New Eugenics and the End of Liberalism

It should be clear, then, that this debate mainly revolves around the question of whether the new eugenics concurs with or contradicts the central principles of liberalism. Broadly speaking, two of these principles seem especially relevant in this context: value pluralism and individual freedom. One of the most striking features of this debate, however, is that while all participants seem equally committed to these principles, they ultimately evaluate the new eugenics in diametrically opposed ways. In order to gain a clearer understanding of how this is possible, it may be helpful to point out that much depends on the actor that is given priority in this debate: while liberal eugenicists tend to emphasize the parent, critics believe that special consideration should be given to the prospective child. This difference of emphasis explains why the former group considers state neutrality in the domain of reproduction to be a sufficient guarantee of the liberality of the new eugenics. They argue that if the state remains neutral in this matter and does not intervene to enforce a particular conception of the good to be sought through genetic modification, then, by giving parents more control over which genetic traits their children will inherit, a liberal eugenics will actually strengthen the freedoms associated with reproduction. Critics, on the other hand, point out that the main threat to the liberal principles of value pluralism and individual freedom no longer comes from potential state intervention but from parents themselves. According to this group, it is not the freedom of parents that is at issue but the freedom of the children born to them. In particular, they stress the fact that children who are genetically designed according to the developmental goals of their parents will be unable to experience themselves as the autonomous authors of their own lives. To put it as succinctly as possible, the liberty of the parent is thus the bondage of the child.

Yet, on the face of it, this way of framing the new eugenics debate may not be entirely satisfactory, for it gives the impression that liberal eugenicists believe there should be no moral or legal limitations whatsoever to the reproductive liberty of parents. This is obviously not the case. As with other individual liberties, reproductive choices tend to be judged for acceptability against John Stuart Mill's principle of harm. As is well known, this principle broadly states that one is free to act as one chooses, as long as one's actions do not cause harm to others.[60] While this immediately raises the question of what counts as harm, there is still a more urgent issue to be considered here: in the case of reproductive liberty, the limit of individual freedom is not set by potential harm done to fellow citizens

[60] John Stuart Mill, *On liberty* (Cambridge: Cambridge University Press, 1989).

but to human beings who do not yet exist. One of the most influential approaches to this complicated issue was developed by Derek Parfit, and is known as the "nonidentity argument."[61] The example Parfit gives is that of a 14-year-old girl who decides to have a child. Intuitively, we would be inclined to believe that she is likely to harm her child because, by dint of having such a young mother, the child is likely to receive "a bad start in life."[62] Furthermore, we would probably also believe that it would have been better for her child if the mother had waited longer to conceive, for then her child would have had better chances in life. Parfit shows, however, that this is an inaccurate appraisal of the situation. If the girl had indeed waited longer to have a child, this child would have been the product of a different egg and a different sperm. It would, in other words, have been a different child. The further implication of this is that the child born to her at the age of 14 has not been harmed, since the condition of this particular child should be compared not to that of the hypothetical child born a couple of years later but to the condition of *not being born at all*. In other words, being born to a 14-year-old mother is no worse for a child than being born to, for example, a 24-year-old mother, because the alternative is not being born at all. One of the conclusions that has been drawn from this argument is that in reproductive freedom, the threshold of harm should be set at the point where the child would have been better off not being born. The underlying rationale is that all forms of life which fall short of this threshold constitute a "life not worth living" or a "wrongful life."[63]

Obviously, the problem that some liberal eugenicists have with this argument is not that it would give prospective parents too little reproductive liberty, but that it would give them too much. Indeed, very few are willing to accept the ultimate conclusion to which this argument seems to lead, namely that parents' reproductive liberty should be so wide as to include even the freedom to endow their children with a physical or psychological disability. If this seems too unlikely to be worth considering, we merely have to recall the Duchesneau case mentioned above to see that this is a realistic possibility. Yet, according to the nonidentity argument, a child would usually not be harmed by such an anomalous reproductive choice, for very few cases are likely to arise in which a child would find herself in such terrible conditions that it would have been better for her not to be born at all. It would, for example, be very difficult to maintain that being

[61] Derek Parfit, *Reasons and Persons* (Oxford: Clarendon Press, 1984), 351-379.

[62] *Ibid.*, 358.

[63] See for example Joel Feinberg (1986), 'Wrongful life and the counterfactual element in harming,' *Social Philosophy and Policy* 4 (1): 145-178.

born deaf is worse than not being born at all. In order to escape this conclusion, liberal eugenicists usually fall back on what is called the principle of procreative beneficence, which, in one version, states that parents "should select the child, of the possible children they could have, who is expected to have the best life, or at least as good a life as the others, based on the relevant available information."[64] Broadly, this principle entails that parents are morally required to give their children the best possible genetic endowment. It is clear, however, that this principle is still much too formal to prevent parents from endowing their children with a disability. Consider again the Duchesneau case. The couple did not believe their child would suffer any serious harm as a result of being deaf. On the contrary, they were convinced that this intervention would actually be beneficial to the child's wellbeing, for it would allow her to participate in the rich culture of the deaf community of her parents.

How, then, do liberal eugenicists attempt to resolve this conflict between the principles of reproductive freedom and procreative beneficence in the case of selecting for disability? One solution could be, first, to define disability as a diseased state and subsequently argue that deliberately creating a disabled child constitutes a clear violation of medical deontology. This solution would not be wholly satisfying, though, because it would be necessary to reintroduce an objectivist notion of normality or normal functioning against which a given condition could be assessed. This is a solution that liberal eugenicists wish to avoid at all costs though. John Harris has therefore proposed defining disability as "a condition that someone has a strong rational preference not to be in and one that is moreover in some sense a harmed condition."[65] To determine whether a given condition is a harmed one, he suggests using what he calls the "emergency room test:"

> I have in mind the sort of condition for which if a patient presented with it unconscious in the emergency room of a hospital and the condition could be easily and immediately reversed, but not reversed unless the doctor acts without delay, a doctor would be negligent were she not to attempt reversal.[66]

According to Harris, the main advantage of this conception of a harmed condition is that it is not defined in relation to the state of nonexistence, or to normal functioning, but "relative to possible alternatives."[67] Suppose, he explains, that someone was brought into the hospital with her little

---

64 Julian Savulescu (2001), 'Procreative beneficence: why we should select the best children,' *Bioethics* 15 (5/6): 415.

65 Harris, *Enhancing Evolution*, 91.

66 *Ibid.*

67 *Ibid.*, 92.

finger severed at the first joint and it could be sewn on again. Although it would obviously be absurd to maintain that the missing end joint of this person's little finger meant that her life would be not worth living, there are nonetheless good moral reasons to maintain that the hospital staff would harm the patient by failing to reattach the finger. According to Harris, the same holds true for all other injuries, diseases and disabilities.

Catherine Mills has fiercely criticized this definition of disability, firstly because it neglects the simple fact that "some disabilities are neither irreversible nor removable,"[68] and secondly because it uses the perspective of an "able-bodied person"[69] as the standard against which to evaluate a given condition. Yet, though this criticism may certainly hold true in the present, Mills seems to ignore the fact that Harris develops this argument in relation to genetic modification technologies of the future. What he actually suggests is that *when we have the choice* to have a child either with or without a disability, we have good moral reasons to choose the second option. Another factor that critics have overlooked is that, as genetic science advances, it is likely not only to increase reproductive freedom and the responsibilities that come with it, but also to change the standards against which we seek to measure a harmed condition:

> It is normal now, for example, to be protected against tetanus; the continued provision of such protection is not merely permissive. If the AIDS pandemic continues unabated and the only prospect, or the best prospect, for stemming its advance is the use of gene therapy to insert genes coding for antibodies to AIDS, *I cannot think that it would be coherent to regard making available such therapy as permissive rather than mandatory.*[70]

We cannot think of a stronger argument against deliberately endowing one's children with a disability. What Harris is saying here is that if parents have the power to prevent their child being born with a disability, they should have not merely the freedom to use this power, but the obligation. The implications of this claim, however, are far more radical than they initially appear. Harris's argument is not that the state should intervene to enforce this obligation – in his view, it is a moral obligation we have to our children[71] – but we have no reason to assume that such demands will not be formulated as soon as these technologies become more widely available.

---

68 Catherine Mills, *Futures of Reproduction. Bioethics and Biopolitics* (Dordrecht: Springer, 2011), 22.

69 *Ibid.*

70 Harris, *Enhancing Evolution*, 93; Emphasis added.

71 See also Julian Savulescu (2009), 'The moral obligation to create children with the best chance of the best life,' *Bioethics* 23 (5): 274-290.

If, upon closer examination, the liberal eugenicists' argument for the *freedom* to intervene in the genetic make-up of future generations resembles an argument for the *obligation* to intervene, then it is seems that the critics are quite right to conclude that, far from concurring with the principles of liberalism, "liberal eugenics is a betrayal of liberal philosophy."[72] Curiously, this is not how they themselves reach this conclusion. As we have seen, the danger that many see in a liberal eugenics is that it might change human nature. Although this argument takes many different forms, the basic idea is that genetically enhanced humans will be so different from their unenhanced predecessors that they will no longer be able to recognize each other as equal members of a free society. Habermas has developed what is probably the most sophisticated version of this approach. His main point of critique is that being endowed with specific genetic traits and characteristics will deprive the programmed person of "an unobstructed future of his own."[73] The idea is that a person who learns that some of her talents, skills and abilities were not given to her by "nature" but by means of the deliberate intervention of another person will find it impossible to understand herself as the singular author of her own life. This argument can be understood in several different ways, but one of the most important seems to be that the programmed person may consider her life to have been steered in a particular direction at the wish of another person.

Interestingly, however, the underlying idea of this argument did not originate in the context of a discussion about the consequences of new genetic technologies. What actually prompted Joel Feinberg to write his seminal essay 'The Child's Right to an Open Future'[74] was a series of lawsuits in which members of the Amish community challenged compulsory schooling laws in various states of the USA. As is widely known, the Amish live an extremely secluded life, far removed from the complexity of the modern industrialized world. Amish communities traditionally educate their children for a life of industry and piety, teaching them only the crafts and skills deemed necessary to carry on their religious lifestyle. External influences are thereby kept to an absolute minimum. In *Wisconsin v. Yoder*[75] the United States Supreme Court ruled in favor of an appeal made by the Amish community, noting that by forcing Amish children to attend state schools the State of Wisconsin infringed on their

72 Dov Fox (2007), 'The Illiberality of "liberal eugenics",' *Ratio* 10 (1): 24.

73 Habermas, *The Future of Human Nature*, 63.

74 Joel Feinberg, 'The Child's Right to an Open Future,' *in Whose Child? Children's Rights, Parental Authority, and State Power*, ed. W. Aiken and H. LaFollette (Totowa, NJ: Rowan & Littlefield, 1980).

75 *Wisconsin v. Yoder*, 406 U.S. 205 (1972).

constitutional religious rights. In his essay, Feinberg disagrees with this decision by arguing that:

> [A]n education that renders a child fit only for one way of life forecloses irrevocably his other options. He may become a pious Amish farmer, but it will difficult to the point of practical impossibility for him to become an engineer, a physician, a research scientist, a lawyer, or a business executive. The chances are good that [...] critical life-decisions will have been made irreversibly for a person well before he reaches the age of full discretion when he should be expected, in a free society, to make them himself.[76]

In other words, according to Feinberg, the Amish way of life infringes on Amish children's right to an open future by prematurely closing off many of the other ways of life available in a free liberal society.

It seems somewhat odd that Habermas refers directly to Feinberg's essay in his argument against enhancement technologies,[77] for two reasons. Firstly, by likening attempts to shape children by altering their social environments to attempts to shape them by altering their genetic profiles, he actually seems to be pursuing one of the strategies that liberal eugenicists employ to argue the opposite of what Habermas himself intends. As explained above, if there are no substantial differences between genetic intervention and the other influences that parents have over the development of their children, then there is no reason to allow the latter while rejecting the former. Secondly, at the core of Habermas's argument lies the contention that while the effects of "a pathogenic socialization process" can always be "revised by critical reappraisal,"[78] this is impossible in cases of genetic intervention. If, as it appears, he actually disagrees with Feinberg's view on the intrusiveness of certain educational practices, why then does he claim to base his own argument on it?

The most plausible explanation for this confusion seems to be that Habermas wishes to retain the structure of Feinberg's reasoning but not its content. That is to say, he agrees with him insofar as we should be especially concerned about a child's right to an open future, but disagrees with him insofar as he rejects the notion that the greatest threat to this right comes from a "pathological" socialization process. Liberal eugenicists often liken the effects of socialization to those of genetic intervention in order to argue that the latter is no more problematic than practices that are now routinely accepted as part of normal parenting. Habermas would be unlikely to disagree with the argument that parents' reproductive freedom

76 Feinberg, *Right to an Open Future*, 132-133.

77 See Habermas, *The Future of Human Nature*, 124.

78 *Ibid.*, 62.

should also encompass genetic interventions. As soon as priority is given to the perspective of the "passive receiver," however, then a very different picture emerges. After all, whereas socialization occurs at a moment when a child is already able to respond to the actions of her educators, genetic intervention occurs before the child has even entered into existence and the resulting individual will therefore be unable to respond effectively to his or her producer's intentions:

> [...] such an imposition from within the community, even if it is excluded from the relationships obtaining between morally acting persons, must nevertheless not be confused with an external or alien determination of the natural and mental constitution of a future person, *prior to an entry into the moral community.*[79]

Most critics revert to the notion of human dignity in order to oppose these kinds of interventions. As we have seen, however, this road is not open to Habermas because it entails giving full rights to unborn life and in his view, it is clear that the question of whether "the in vitro embryo were already 'another,' who possessed completely valid basic rights [...] can hardly be answered in the positive given the premises of an ethically neutral constitutional order."[80] If it is already extremely difficult – if not impossible – to reach consensus on the question of when life begins, then these problems are only likely to increase in the case of genetic intervention, for gene modification can be performed not only at the zygote and embryo stages, but also in sperm and egg cells. It seem quite reasonable to assume that few would be willing to accept the absurd consequences that would follow from giving sperm and egg cells full human rights. While the proposed dilemma is quite clear, however, the same cannot be said about Habermas's solution to it. When he contends that "legal protection might come to be expressed in a right to a genetic inheritance immune from artificial intervention,"[81] then it remains far from evident who might be the beneficiary of this right. Since he rejects the idea of giving such a right to prenatal forms of life, he seems to mean that it would be bestowed upon the adult enhanced person. But how could such a person ever exercise her right to a genetic endowment free from artificial intervention, given that this irreversible act would have taken place well before she was a position to do so?

There is more to be said here, though. For what the debate between the advocates and opponents of a liberal eugenics also makes evident is

---

79 *Ibid.*, 79; Emphasis added.

80 *Ibid.*, 77.

81 *Ibid.*, 27.

that the emergence of enhancement technologies is likely to be accompanied by a growing tendency to impose severe normative constraints on certain potential forms of life. This is clear enough in the arguments of Fukuyama and Habermas, who draw on a normative conception of human nature to argue against genetic enhancement. What has not been sufficiently emphasized thus far, however, is the fact that any attempt to give normative content to human nature may be mobilized politically to exclude those who deviate from this norm.[82] That is not to say that these authors' conceptions of human nature could serve as grounds for excluding certain vulnerable groups, such as the disabled or the mentally ill, from the moral community. Instead, it could be said that these definitions preemptively deny any genetically enhanced being that may be brought into existence in the future the status of human being. What else could Habermas mean when he writes that "[t]his new type of relationship [between programming parent and programmed child, NVC] offends our moral sensibility because it constitutes *a foreign body* in the legally institutionalized relations of recognition in modern societies"[83]? This statement seems, moreover, to cast further doubt on the effectiveness of Habermas's call for a right to a genetic constitution free from genetic intervention. If an enhanced person is barred from establishing reciprocal relationships with "normal" human beings, and thus from entering the moral community of equal citizens, on what grounds, then, may such a person appeal to this right in the first place? Again we must ask who the bearers of this right would be if the only individuals to have an interest in it were denied legal subjectivity?

This tendency is not absent from the discourse of liberal eugenics, however. Quite the contrary, in fact. As explained above, many liberal eugenicists seek to avoid some of the more distressing consequences of the nonidentity argument by tempering the right to reproductive liberty with the principle of procreative beneficence. In their view, this is especially urgent as new genetic technologies continue to emerge, because these technologies will increase parents' freedom to modify their offspring not only for the better, but also for the worse. Furthermore, we have shown that by insisting on the principle of procreative beneficence, liberal eugenicists actually argue for the obligation, rather than simply the freedom, to endow future generations with the best possible genes. The ultimate result of this argument is that parents would have the obligation both to prevent their children from being born with a disability or with a disease

82 See for example Eduardo Mendieta (2003), 'Communicative freedom and genetic engineering,' *Logos* 2 (1): 135-138.

83 Habermas, *The Future of Human Nature*, 14; Emphasis added.

and to boost their capacities to a maximum. We should not lose sight of the justification behind this line of reasoning, however. What liberal eugenicists reject is not the notion of "wrongful life" as such, but only the criteria which are to be used to determine what forms of life are included in this category once genetic technologies become available. What they are actually arguing, therefore, is that while it may be true that it is currently better, for example, to be born deaf than not born at all, this may change once we have the power to choose between a deaf child and a hearing child. If it is true, on the other hand, that the emergence of genetic technologies will progressively raise the threshold of harm, then we are also about to witness a steady increase in the number of forms of life that will have to be categorized as wrongful life. It remains to be seen how far this category can be stretched but perhaps, in the not too distant future, human beings as we currently know them will all be judged as having a "life not worth living."

Human genetic modification is still in its infancy, but the issues discussed above suggest that liberal political and moral philosophy remains rather ill equipped to address this controversial field. This is evident in the mere fact that a commitment to the central principles of liberalism is able to produce two diametrically opposed views on the subject, and also in that the two positions appear to be conceptually inconsistent: ultimately, both lead to conspicuously illiberal conclusions. After all, as we have argued above, while the argument against a new eugenics necessarily entails a preemptive dehumanization of any enhanced form of life, the argument for it threatens to reduce any non-enhanced form of life to the status of wrongful life. The final analysis might conclude, then, that any kind of liberal response to the challenges of the new eugenics unwittingly produces a form of life devoid of any intrinsic value. This is not to say that this outcome is inevitable. Clearly, though, we will need to rely on an alternate interpretative framework if we wish to gain a more precise understanding of the issues involved, since we appear to have left that of liberal political philosophy and unwittingly entered that of *biopolitics*.

Chapter Two

Bare Life

> [A]t the opposite extreme, you no longer have a sovereign right that is in excess of biopower, but a biopower that is in excess of sovereign right. This excess of biopower appears when it becomes technologically and politically possible for man not only to manage life but to make it proliferate, to create living matter, to build the monster, and, ultimately, to build viruses that cannot be controlled and that are universally destructive. This formidable extension of biopower [...] will put it beyond all human sovereignty.[1]

The Biopolitical Turn

Recently, the notion of biopolitics has begun to receive increasing attention in critical political thought, and is rapidly establishing itself as an independent theoretical paradigm. One of the main reasons why this notion has seen such a sudden growth in popularity, and has even found its way into such diverse fields as bioethics, feminist studies and the social sciences, is that it purports to explain a large variety of highly relevant contemporary social and political events which seem to slip through the grasp of more traditional theoretical interpretations. Biopolitics can indeed be linked to issues as seemingly unrelated as birth control policies, ethnic conflicts, the proliferation of surveillance cameras in the public domain, unlawful detainee camps, the ramifications of climate change, the fight against illegal immigration, the threat of pandemics, "humanitarian" wars, and even agricultural subsidy policies. Yet, as diverse as these phenomena may seem, they all draw attention to the fact that the contemporary world is experiencing an increased politicization of human beings' biological existence. If standard political theory is unable to produce a satisfying explanation of what is at stake in this process, it is mainly because its conceptual framework is not adequate for the task at hand. Throughout the history of political thought, virtually all major commentators have defined political action in terms of how it differs from activities driven by biological wants and needs. Politics, in the classic sense, entails free human interaction and performing actions that are worth pursuing as ends in themselves, and therefore transcends the realm

---

1 Michel Foucault, *Society Must Be Defended: Lectures at the Collège de France, 1975-76*, trans. D. Macey (London: Penguin Books, 2003), 253-254.

of biological necessity. Thus, from this point of view, a politics concerned purely with human biological life can only be a contradiction in terms.

It is certainly not the case, however, that this change went entirely unnoticed in earlier political thought. When, for example, Hannah Arendt referred to "the rise of the social" and the concomitant "victory of the *animal laborans*"[2] as the single most important event of political modernity, she placed herself firmly within a biopolitical framework. The fact that her penetrating analysis received severe criticism for projecting a romanticized glorification of Greek *polis* life onto modern industrialized society is, in that sense, just one sign that political theory was not yet ready to rethink its basic assumptions and situate this epochal event in its proper context.[3] Arendt herself was, of course, very much aware that she was entering uncharted territory, and that drawing extensively on the Greek *polis* experience was bound to lead to serious misinterpretations. Arguably, if Arendt thought it necessary to deal at great length with the original Greek understanding of politics in *The Human Condition*, it is because here we can still catch a glimpse of the basic political categories in their original, undistorted forms. This is not the most difficult aspect of her work to understand, however, for the real problem arises when she attempts to explain the meaning and sense of these categories in an entirely new context. She found it difficult not only to determine the precise nature of the central political subject once the boundary between the private realm (*oikos*) and the public realm (*polis*) ceases to exist, but also to conceive of the nature of the relation between those two realms once they begin to overlap. As for the first difficulty, the *homo laborans* could certainly not be a subject of free will. But neither was she drawing on the Greek *bíos politikos*, for although the latter term contains a clear reference to the sphere of life in which the *homo laborans* was traditionally situated, in Greek understanding this term denotes precisely the political existence of human beings in the public sphere. If anything, Arendt seems to be referring to *zoē*: human life in its most elemental biological dimension. How, though, are we to conceive of a politics constructed around the most apolitical mode of existence imaginable? As for the second difficulty, we should consider what it means exactly to say that the boundaries between the private and the public are blurred. Does it mean that the ends of politics are now dictated solely by the biological necessities and limitations of human beings? Or that politics now seeks

2 Hannah Arendt, *The Human Condition* (Chicago: The University of Chicago Press, 1998), 320.

3 See for example Martin Jay (1978), 'The Political Existentialism of Hannah Arendt,' *Partisan Review* 45 (3): 348-369 and Sheldon Wolin (1983), 'Democracy & The Political,' *Salmagundi* 60: 3-19.

to extend its reach to include human biological life within its sphere of power? In other words, how are we to understand the relation between life and politics?

While these are some of the questions central to the field of biopolitics, Arendt was not the first author who sought to address them; nor did she ever attempt to explore systematically the specific context from which they emerged, at least not explicitly. In order to gain a clearer understanding of the tension between the two concepts present in the notion of biopolitics, namely a politics *of* life and politics *over* life, we will start this chapter with a brief overview of the field's history and the changes it has undergone.[4] Subsequently, we will turn to Foucault's theory of biopolitics, for it is in his writings that the tension mentioned above is most clearly articulated. Foucault formulates this tension as a paradox: How is it possible that a form of power that seeks to protect and strengthen life was also able to generate a political regime – Nazism – whose eugenic practices aimed to suppress life? In the recent literature, two solutions have been suggested for resolving this conflict: Roberto Esposito's paradigm of immunity and Giorgio Agamben's theory of the state of exception. After discussing the basic tenets of these solutions, I will suggest that they can also throw fresh light on the debate surrounding genetic intervention.

From Life Politics to a Politics of Life

The first author to use the term biopolitics is likely to have been the Swedish political scientist Rudolph Kjellén, in a book with the telling title *The State as Form of Life* (1916).[5] In an earlier work, Kjellén had already shown that dynamic states with limited territory and scarce natural resources are compelled to conquer weaker states in order to secure sufficient "vital space" (*Lebensraum*) for their ever-growing populations.[6] In *The State as Form of Life*, he subsequently argued that such "geopolitical" behavior, as he insisted on calling it, can only be explained if one assumes that states are not the artificial creations of free consenting

4 For a more comprehensive historical overview of the notion of biopolitics, see Thomas Lemke, *Biopolitics: An Advanced Introduction*, trans. E. F. Trump (New York and London: New York University Press, 2011), 9-32 and Roberto Esposito, *Bíos: Biopolitics and Philosophy*, trans. T. Campbell (Minneapolis: The University of Minnesota Press, 2008), 16-24.

5 Rudolph Kjellén, *Der Staat als Lebensform* (Berlin: Kurt Vowinckel Verlag, 1924). Originally published in Swedish as *Staten som Lifsform* (Stockholm: Hugo Geber, 1916).

6 Rudolph Kjellén, *Grundriss zu einen System der Politik* (Leipzig: Rudolf Leipzig Hirzel, 1920).

individuals, as the liberal contractualist tradition had always maintained, but in fact "living organisms" whose functioning transcends the will of those belonging to them. In Kjellén's view, the different groups and classes in society represent nothing but the state-body's organs, which incessantly strive to preserve themselves and to increase their power. However, in times of extreme crisis or external war, they feel the need for cooperation in order to secure the existence of the state organism as a whole. Kjellén's organic conception of the state was to become especially popular in Weimar Germany, where scholars were already familiar with biologistic views of reality because of the influence of the *Lebensphilosophie* tradition. Moreover, the fact that Weimar Germany was going through a series of major political crises, marked by mass strikes and the threat of civil war, certainly played a role in how the organic theory of the state was received here. One only has to look at Jacob von Uexküll's influential book *Staatsbiologie* (1920)[7] to see that Weimar intellectuals were interested in this theory not only because of its explanatory value, but also because of the solution it offered. In von Uexküll's view, translating political phenomena into biological terms was only the first step to be taken. In addition to this, it was also deemed necessary to introduce the language of medicine, for it is one thing to diagnose the diseases threatening the existence of the state organism – such as electoral democracy and the right to strike – but quite another to propose a remedy for them. He therefore proposed to create "a class of state doctors"[8] who would be in charge of detecting and removing the tumors that were spreading throughout the body of the state organism.

From here it was only a small step to the National Socialists' idea of the state as a racially homogeneous *Volkskörper* ("national body"). Central to the National Socialist ideology was the assumption that the state was not composed of an aggregate of individuals, but that it was an organic community with a shared biological heritage. In contrast to its liberal and communist competitors, this ideology did not see history as a continuous struggle between interest-driven individuals or between competing social classes, but between different races. Moreover, it assumed that there was a natural inequality between the races, formed primarily by hereditary factors, such that it was possible to classify human beings according to a strict hierarchical order. National Socialist ideologues not only claimed that this racial worldview was based on objective scientific evidence, they also believed that politics should be arranged accordingly.

---

7 Jacob von Uexküll, *Staatsbiologie: Anatomie, Physiologie, Pathologie des Staates* (Berlin: Verlag von Gebrüder Paetel, 1920).

8 *Ibid.*, 55.

In the words of Hans Reiter, one of the leading physicians of the *Reich*, "[i]t is inevitable that this course of thought should lead to the recognition of biological thinking as the baseline, direction, and substructure of every effective politics."[9]

The problem that faced Germany and the entire West was an ever-worsening process of bio-spiritual degeneration, which threatened to enfeeble the national body and ultimately cause it to disintegrate. Yet, although the National Socialists believed that the mythical forces of race and heredity were beyond human control, they were also convinced that it was still possible to counter this degenerative tendency and restore the national body to health and perfection. To accomplish this, they implemented a series of politico-medical interventions on the national body, carried out in the name of "racial hygiene" (*Rassenhygiene*). For the most part, works on racial hygiene, such as those by Reiter,[10] Verschuer,[11] and Ploetz,[12] advocated the use of both positive and negative eugenics to regenerate the race. In an attempt to increase the number of true Aryans, the regime launched a staunch pro-birth campaign. In addition to this, it also founded the *Lebensborn* project, which sought to produce a superior race by means of selective breeding. More attention was paid to negative eugenic measures, however. Since it would take several generations for the results of a positive eugenics to become visible, in the short term, it would be necessary to hinder the reproduction of those endowed with what were deemed to be inferior hereditary traits. Measures along these lines were implemented almost as soon as the Nazi regime came to power. As early as July 1933, the 'Law for the Prevention of Hereditarily Diseased Offspring' was adopted, which ordered the sterilization of people affected by conditions which were thought to be hereditary, such as imbecility, epilepsy, manic depression, blindness and even alcoholism. Moreover, to prevent interracial mixing, the 'Law for the Protection of German Blood and German Honor' was passed, prohibiting marriages between Aryan Germans and Jews. But these measures were only a prelude to what was to come. In 1939, the regime launched the T4 program, which authorized the use of euthanasia among the genetically "unfit," and

9 Hans Reiter, 'Unsere Biopolitik und das Auslandsdeutschtum,' in *Das Reichsgesundheitsamt 1933-1939: Sechs Jahre Nationalsozialistische Führung* (Berlin: Julius Springer Verlag, 1939), 37-43.

10 Hans Reiter and Johannes Breger (eds.), *Deutsches Gold: Gesundes Leben – Frohes Schaffen* (München: Röhrig, 1942).

11 Otmar von Verschuer, *Leitfaden der Rassenhygiene* (Leipzig: Thieme, 1941).

12 Alfred Ploetz, *Die Tüchtigkeit unserer Rasse und der Schutz der Schwachen* (Berlin: Fischer Verlag, 1895).

in an ultimate attempt to cleanse the national body of harmful elements, it finally ordered their total destruction.

While the biopolitical dream of literally transforming politics into a form of applied biology was abandoned after the Second World War, the basic theoretical framework upon which it was based remained more or less intact. Postwar political scientists who continue to study political phenomena from a naturalistic perspective have taken great pains to point out that this understanding of biopolitics does not necessarily lead one to prefer one particular political system over another, but their assumptions are not radically different from their predecessors. Advocates of a naturalist approach to politics, which remains mainly an American affair, argue that the methodology of the social sciences is fundamentally flawed insofar as it ignores the import of biological factors in its analyses.[13] In particular, they reject the idea that the subjective motives and intentions of political actors are key to understanding political phenomena, and propose that attention should be directed towards these actors' empirically observable behavior. In the literature, two main approaches can be discerned.[14] The first of these draws on sociobiological and ethological research to understand the biogenetic causes of political behavior. It argues that most of these causes, such as competition, cooperation and aggression, are interconnected with genetically programmed dispositions. If these behaviors can be observed universally, it is because they are essential to survival and thus "evolutionarily adaptive." The second approach takes its inspiration from Neo-Darwinian evolutionary theory in describing and explaining the origins of state structures. Most authors working in this area agree that authoritarianism conforms most closely to human nature, since this form of government is the one most likely to result from genetically based behaviors that are central to human sociability, such as territorial drive and hierarchy formation. In this view, the emergence of democracy is not an evolutionary aberration, but it is possible only under very specific and rather rare evolutionary conditions and is therefore bound to remain an exceptional event.[15] That is not to say, however, that advocates of this naturalist approach are

13 See for example James C. Davies, *Human Nature in Politics: The Dynamics of Political Behavior* (New York: Wiley, 1963), Roger D. Masters, *The Nature of Politics* (New Haven: Yale University Press, 1989), and Albert Somit and Steven A. Peterson, *Darwinism, Dominance, and Democracy: The Biological Bases of Authoritarianism* (Westport: Praeger, 1997).

14 See Robert H. Blank and Samuel M. Hines, *Biology and Political Science* (New York: Routledge, 2001), esp. 79-98.

15 Tatu Vanhagen, *Prospects of Democracy: A Study of 172 Countries* (New York: Routledge, 1997), 21-26.

interested in formulating a model of politics directed at the preservation and promotion of the evolutionary conditions favorable to the democratic spirit. The conclusion they draw from their findings is that the only politics realistically possible is one that conforms to the immutable laws of biology.

Although the naturalist variant of biopolitics has not completely disappeared from political science, it has been gradually superseded by a different approach to the question of the relation between life and politics. In simple terms, we might say that this shift in focus was necessitated by the recognition that politics is not merely the passive outcome of bio-natural processes, but that politics also actively intervenes in these processes. This new understanding of biopolitics has two different versions, each triggered by a particular set of events. The first version concerns itself with the detrimental effects that human activities have on the natural environment.[16] Since the 1960s, we have seen a growing awareness that unlimited economic growth threatens to destroy the biosphere on which humanity depends for its survival. It was predicted that the worsening of environmental problems, such as air and water pollution, the depletion of natural resources, species extinction and population explosion, would cause severe economic and social problems, and could eventually lead to a global political crisis and even to war. In order to tackle these problems, a comprehensive program of socio-political measures was proposed which primarily sought to protect the environment from excessive human activities. During the early 1990s, when the issue of climate change became a global political concern, this approach was supplemented by policies built on the assumption that the most effective means of addressing environmental problems was not to change attitudes and behaviors, but to develop "environmentally friendly" or "green" technologies.

The second version of a biopolitics premised on the idea that politics increasingly involves the regulation and organization of natural processes is also concerned with human intervention in nature but, in contrast to the ecopolitical variant, it focuses not on interventions in the human's "outer nature" but on interventions in its "inner nature." This strand of biopolitics is especially important for our topic, because it was conceived in direct response to the proliferation of new biotechnologies. The main idea guiding this approach is that if the development of genetic technologies demonstrates that there are no impermeable boundaries between nature and culture, then renewed political and legal efforts must be made

16 For a general introduction to this field, see Neil Carter, *The Politics of the Environment: Ideas, Activism, Policy* (Cambridge: Cambridge University Press, 2001).

to retrace and reestablish these boundaries.[17] As we saw in the preceding chapter, this biopolitical strategy can take a variety of forms. The most prevalent of these strategies seeks to limit technological intervention in the human genome by constructing a robust moral conception of human nature. Advocates of this approach argue that a substantive idea of human nature is intrinsic to our conceptions of justice and human rights. Genetic interventions are therefore seen as a threat to our moral and political self-understanding as free acting agents. Liberal eugenicists, on the other hand, pursue a diametrically opposed strategy. They argue that, far from contradicting the freedoms and rights associated with liberalism, genetic technologies actually promise to extend them significantly.

We have argued that both positions are highly problematic, because both involve the dehumanization of a particular form of human biological life. This is a puzzling outcome, especially in view of the fact that, while the advocates of a liberal eugenics aim "to create healthier, longer-lived, and altogether 'better' individuals,"[18] its detractors seek to preserve the biological foundations of human existence. For how is it possible that these attempts to conceive of a life-centered and life-affirming politics become transformed into the opposite of what they intend? At this point, we can see that something remains unthematized in the classic notion of biopolitics, something that impinges directly on the relation between the two central concepts under discussion. Thomas Lemke makes the point with particular clarity when he states that

> [t]he interpretation of biopolitics as a mere province of traditional politics is inadequate, in that it presumes that the substance of the political sphere remains untouched by the growing technological possibilities for regulating life processes. This, however, is not the case. Biopolitical questions are fundamental precisely because not only are they objects of political discourse, but they also encompass the political subject him- or herself.[19]

In other words, despite clear differences, the two positions presuppose the existence of a stable, autonomous subject, both an individual who is capable of making political decisions concerning the boundary between the "natural" and the "artificial," the "grown" and the "made," and an individual for whom such decisions are to be made. The problem, however, is that the nature of this subjectivity is less an origin than an effect of political decisions. In this view, biopolitics is less a particular kind of

[17] Odelia Funke (1985), 'Biopolitics and public policy: controlling biotechnology,' *Political Science* 18 (1): 69-77.

[18] Harris, *Enhancing Evolution*, 5.

[19] Lemke, *Biopolitics: An Advanced Introduction*, 30.

politics that deals with issues related to the biological existence of human beings than it is a force field in which the meaning, value and purpose of these notions are configured. Perhaps nobody saw this more clearly than Michel Foucault.

The Power to "Make Life" and "Let Die"

Given the fact that Foucault only occasionally mentioned the notion of biopolitics explicitly in his writings, it may be somewhat surprising that he is now generally referred to as the most important author in this field, or even its founding father. The term "biopolitics" itself makes its first appearance in a short lecture he gave at a conference in Rio de Janeiro in 1974,[20] and then returns in the final session of his 1976 Collège de France lectures,[21] and again at the very end of the first volume of *The History of Sexuality*,[22] which appeared the same year. In addition, while the titles of his 1978 and 1979 Collège de France lectures give the impression that he intended to flesh out the notion of biopolitics more thoroughly here, one only need look at the topics discussed in these lectures to conclude that they appear to deal with a somewhat different problematic, most notably that of governmentality.[23] However, if this appears to suggest that the issue of biopolitics was only of minor importance to Foucault, we should look again; in reality, it was central to his work at least from the publication of *Discipline and Punish* on.[24] The latter book is presented as a historical inquiry into the profound changes undergone by the penal system over the previous three centuries. But the evolution described here, from a system centered on public executions to one based on closely monitored prisons, is but one manifestation of a much more fundamental transformation in the power structures of modern societies. Foucault saw this shift as entailing a transformation in the mechanisms of power away from that of sovereignty, which deals primarily with legal subjects, to that of biopower, which focuses on the biological existence of human beings:

---

20 Michel Foucault (2004), 'The crisis of Medicine or the crisis of antimedicine?' *Foucault Studies* 1: 5-19.

21 Foucault, *Society Must Be Defended*, 239-264.

22 Michel Foucault, *The History of Sexuality, Vol. 1: The Will to Knowledge* (New York: Penguin Books, 1998).

23 Michel Foucault, *Security, Territory, Population: Lectures at the Collège de France 1977-1978*, trans. G. Burchell (Basingstoke: Macmillan, 2007), Michel Foucault, *The Birth of Biopolitics: Lectures at the Collège de France 1978-1979*, trans. G. Burchell (Basingstoke: Palgrave Macmillan, 2008).

24 Michel Foucault, *Discipline and Punish: The Birth of the Prison* (New York: Pantheon Books, 1977).

> It seems to me that one of the basic phenomena of the nineteenth century was what might be called power's hold over life. What I mean is the acquisition of power over man insofar as man is a living being, that the biological came under State control, that there was at least a certain tendency that leads to what might be termed State control over the biological.[25]

In *The History of Sexuality*, Foucault explains the principal difference between these two regimes of power in terms of the contrast between "deduction" and "production." According to him, sovereign power operates through the "right to seizure: of things, time, bodies, and ultimately life itself."[26] It reserves the right to deduce or to extract goods and services from its subjects. The most extreme manifestation of this sovereign right is the capacity to decide on the life and death of subjects. From the perspective of the sovereign, life and death are not natural phenomena but constitute the outer limits of the domain which falls within its sphere of power. Only the sovereign can grant the subject the right to be alive or to die. However, in practice, priority is granted to the sovereign's right to kill: the sovereign can exercise its right over life only through its decision to kill or not to kill.

During the classical age, this "ancient right to *take* life or *let* live" was gradually replaced by "a power to *foster* life or *disallow* it to the point of death."[27] In contrast to sovereign power, this new power mechanism does not function by suppressing life and sapping its strength, but seeks to make it productive by controlling, regulating, reinforcing and optimizing the vital processes pertaining to it. It is impossible to pinpoint exactly what caused this fundamental transformation in the power structure, but it would certainly not have been possible without a number of developments that occurred outside of the field of politics, such as the steady increase in economic and agricultural productivity and the availability of new scientific and medical knowledge. As a result of these developments, the biological existence of human beings was no longer considered to be the impassable natural limit of politics, but gradually became its prime object:

> For the first time in history, no doubt, biological existence was reflected in political existence; the fact of living was no longer an inaccessible substrate that only emerged from time to time, amid the randomness of death and its fatality; part of it passed into knowledge's field of control and power's sphere of intervention.[28]

25 Foucault, *Society Must Be Defended*, 239-240.

26 Foucault, *The History of Sexuality*, 136.

27 *Ibid.*, 138.

28 *Ibid.*, 142.

That is not to say, however, that it would ever be possible for this power mechanism to control life in all its forms. As soon as life is no longer regarded as an unalterable given, and enters the realm of explicit political calculations, hitherto unknown dimensions of life show themselves everywhere. Seen from this perspective, life and politics are engaged in a never-ending struggle in which neither side is ever strong enough to completely master the other. For Foucault, then, biopolitics is situated precisely in this "dual position of life that placed it at the same time outside history, in its biological environment, and inside human historicity, penetrated by the latter's techniques of knowledge and power."[29]

Historically and conceptually, Foucault distinguishes between two dimensions of this life-oriented politics: an "anatomo-politics of the body" and a "bio-politics of the population."[30] The first of these emerged in the late 17th century and centered on the individual body as a machine. It mainly sought to create politically docile and economically productive individuals by subjecting their bodies to the supervision and control of disciplinary technologies in various institutions such as schools, army barracks, asylums, hospitals, factories and prisons. In the second half of the 18th century, this disciplinary power was complemented by a new power technology, one which no longer targeted "man-as-body" but "man-as-living-being."[31] This "security technology,"[32] as Foucault called it, was not only employed in a much more centralized manner, it also operated on a larger scale. It included, for example, the development of state policy in relation to birth control and the problem of morbidity. In order to address the challenges and dangers bound up with these phenomena, a system of health care was set up which mainly focused on public hygiene. Other important fields of intervention included those related to the problems associated with old age, accidents, infirmities and various anomalies. This led to the creation of new control mechanisms such as insurance, collective savings and safety measures. Furthermore, the biopolitical regime also began to focus its attention on conditions related to the natural and artificial environment insofar as these had an influence on the well-being of those living in them: the draining of swamps, the building of dikes, urban development, and so on. What all these phenomena have in common is, firstly, that they are "collective phenomena which have their economic and political effects,"[33] and that they are visible only

---

29 *Ibid.*, 143.

30 *Ibid.*, 139.

31 Foucault, *Society Must Be Defended*, 242.

32 *Ibid.*, 249.

33 *Ibid.*, 246.

at population level. In other words, the risks and dangers inherent to these phenomena all result from the existence of a population as a biological entity. Secondly, they are all essentially "aleatory events that occur within a population that exists over a period of time."[34] This means that biopolitics will employ mechanisms with very different functions from those employed by disciplinary power: forecasts, statistical estimates, and so on. These mechanisms do not intervene at the level of the individual but at the level of the population, and they do this in such a way as to achieve a general equilibrium: "it is, in a word, a matter of taking control of life and the biological processes of man-as-species and of ensuring that they are not disciplined, but regularized."[35]

In the 1976 Collège de France lectures and in *The History of Sexuality*, Foucault's outline of the concept of biopolitics remains somewhat schematic; it is analyzed in more detail, however, in the 1978 and 1979 lectures. In these lectures, he seeks to excavate the genealogy of biopolitics through a historico-critical examination of the process of "governmentalization" as it developed from "the archaic model of the Christian pastorate," through the 17th century "diplomatic-military model," to that of "the police."[36] What this analysis adds to our understanding of biopolitics is that it is not a unidirectional relationship in which life is simply at the mercy of power mechanisms, but that it also produces new forms of subjectivity. Crucially, Foucault does not conceive of government as a purely political concept but places it in a much broader analytical context. In this enlarged sense, government refers not only to the political art of governing men, but also to various religious, philosophical, economic and pedagogical practices of control, guidance and management. The main thread of Foucault's argument is that the modern state is the result of a complex combination of Greco-Roman and Christian forms of government. While the former developed mechanisms for establishing obedience to legal and moral norms, the latter focused on techniques for the comprehensive guidance of individuals. Christian pastoral power centers on the relationship between shepherd and flock and is concerned with the notion that individuals conduct themselves in such a way as to ensure the salvation of their souls in the hereafter. Importantly, the shepherd does not merely seek to enforce his will upon his subjects, but exercises his power by securing knowledge of their inner truth through a variety of methods such as confession:

---

34 *Ibid.*

35 *Ibid.*, 246-247.

36 Foucault, *Security, Territory, Population*, 110.

> [I]f the Christian pastor teaches the truth, if he forces men, the sheep, to accept a certain truth, the Christian pastorate is also absolutely innovative in establishing a structure, a technique of, at once, power, investigation, self-examination, and the examination of others, by which a certain inner truth of the hidden soul, becomes the element through which the pastor's power is exercised, by which obedience is practiced, by which the relationship of complete obedience is assured, and through which, precisely, the economy of merits and faults passes.[37]

What is therefore peculiar about this form of power is that it requires subjects to participate in their own subjugation. But by doing so, by examining and exposing the content of their own consciences, the subjugated also come to see themselves as individual subjects: "[T]he person who examined himself could take control and become master of himself by knowing exactly what he had done and in what respect he had made progress. It was therefore a condition of self-mastery."[38]

According to Foucault, this dialectic between subjugation and subjectivation inherent to the pastoral model of power will have a decisive impact on the formation of the modern state. In the 17th century, it reappeared in secularized form as the doctrine of "reason of state." The defining characteristic of this art of governing was that it no longer relied either on theological principles or on the particular interests of the monarch, but acquired its own proper rationality. What Foucault means by this is that the purpose of government was no longer simply to preserve and increase the power and wealth of the sovereign, but to improve the welfare of the population. That is not to say that governmental power abandoned the ambition to establish dominance over its subjects. But it strove to achieve this precisely by adhering to those subjects' demands and needs. On the one hand, then, governmental power began to take responsibility for the well-being of its population by taking measures to improve living conditions, increase wealth, and so on. On the other hand, however, it was also necessary to extend governmental power to areas of life which had previously been considered part of the private sphere, such as the economy and individual health.

This mutually reinforcing logic was to receive its most comprehensive expression in the 18th century's "science of the police." According to Foucault's usage, the term "police" does not signify the institution of the police in the modern sense, as in "the arm of law," but refers instead to "the set of means by which the state's forces can be increased while

---

37 *Ibid.*, 183.

38 *Ibid.*, 182.

preserving the state in good order."[39] By means of a reading of von Justi's *Elements of Police* (1756) and De Lamare's *Treatise on the Police* (1729), Foucault shows that the aim of policing was to foster all aspects of the lives of the citizens, not only their material and physical conditions but also their moral and spiritual ones. More specifically, the science of police concerned itself with the number of citizens, life necessities such as food and housing, health, work and the circulation of bodies and goods. In other words, this new power mechanism was no longer commanded by "the immediate problem of surviving and not dying, but [was] now commanded by the problem of living and doing a little better than just living."[40] But just as with reason of state thinking, the science of the police did not seek to improve the lives of citizens as an end in itself, but developed interventions and means to "ensure that living, better than just living, coexisting will be effectively useful to the constitution and development of the state's forces."[41]

Foucault's genealogy of governmentality brings the contrast between the paradigms of sovereignty and biopolitics into the sharpest possible relief. While sovereign power was a preeminently negative or deductive power, which suppressed life in order to preserve itself, biopolitics may be said to be a positive or productive power, which fosters and strengthens life in order to extend itself. Yet, if a biopolitical regime can reproduce and expand only by optimizing, reinforcing and multiplying the lives of its subjects, then it is also bound to experience opposition:

> [A]gainst this power that was still new in the nineteenth century, the forces that resisted relied for support on the very thing it invested, that is, on life and man as a living being. [...] [W]hat was demanded and what served as an objective was life, understood as the basic needs, man's concrete essence, the realization of his potential, a plenitude of the possible. [...] Life as a political object was in a sense taken at face value and turned back against the system that was bent on controlling it.[42]

However, according to Foucault, these forms of resistance do not really pose a threat to the biopolitical system as such and are the prelude to a world in which biopolitical conflicts surrounding claims about "the 'right' to life, to one's body, to health, to happiness, to the satisfaction of needs"[43] will become increasingly important.

---

39 *Ibid.*, 313.

40 *Ibid.*, 326.

41 *Ibid.*, 327.

42 Foucault, *The History of Sexuality*, 144-145.

43 *Ibid.*, 145.

What form will these conflicts take and what will be their outcome? Will they lead to the creation of new areas of freedom and result in the emergence of new forms of subjectivity, or will they in fact enable power to tighten its grip on life? If Foucault does not seem to provide a definitive answer to this question, it is because it highlights a paradox in his conception of biopolitics which he was never really able to account for. For how is it possible that the rising to dominance of a form of power that precisely seeks to protect and strengthen life has also generated totalitarian regimes which unleashed a wave of destruction never seen before in history?:

> If it is true that the power of sovereignty is increasingly on the retreat and that the disciplinary or regulatory disciplinary power in on the advance, how will the power to kill and the function of murder operate in this technology of power, which takes life as both its object and its objective? How can a power such as this kill, if it is true that its basic function is to improve life, to prolong its duration, to improve its chances, to avoid accidents, and to compensate for failing? [...] How can the power of death, the function of death, be exercised in a political system centered upon biopower?[44]

What is at stake in this question is the relation between sovereign power and biopower, between "the right to take life" and "the right to make live," or again, between the power *over* life and the power *of* life. If it is true that "Nazism was in fact the paroxysmal development of the new power mechanism," in the sense that "controlling the random element inherent in biological processes was one of the regime's immediate objectives,"[45] how, then, should we understand the fact that it also exercised the sovereign right to kill to a degree hitherto unparalleled, even to the point of complete self-destruction? At this point, Foucault seems to have been left with two possible responses, both of which are clearly unsatisfactory: either totalitarianism marked a temporary break in the otherwise steady evolution from sovereign power to biopower, or it showed that, in a completely enclosed biopolitical order, power is freed from all previously existing boundaries (legal, religious, etc.) and is therefore compelled to discharge itself on the very same vital element from which it draws its strength. Choosing the first option would force Foucault to explain totalitarianism as an anomaly – and thus leave it unexplained – in order to save his theory from being invalidated. In choosing the second, he would be forced to argue that genocidal totalitarianism constitutes the culmination of the entire edifice of political modernity.[46] In

---

44 Foucault, *Society Must Be Defended*, 254.

45 *Ibid.*, 259.

46 Some of Foucault's remarks actually point in that direction: "Of course, Nazism alone took the play between the sovereign right to kill and the mechanisms of biopower to

the end, Foucault attempts to escape this dilemma by arguing that racism provides the point of articulation between sovereign power and biopower in the mechanism of totalitarianism, because racism functions as a means of "establishing a biological caesura within the population [...] between what must live and what must die."[47] However, this explanation has the unfortunate side effect of additionally forcing him to position communist totalitarianism within a racist matrix;[48] not only is this less than convincing, it also threatens to reduce Nazi biopolitics to the status of an example.

It will always remain an open question whether Foucault had really reached an impasse in his analysis of biopolitics or whether his untimely death kept him from developing a theoretical perspective that would enable him to overcome these striking inconsistencies. Be that as it may, in recent literature, two solutions have been suggested to Foucault's paradox of a biopolitics that could simultaneously nurture life and kill, both of which seek to explain the coincidence of these seemingly mutually exclusive poles within a single interpretative framework: Roberto Esposito's paradigm of immunity and Giorgio Agamben's work on the state of exception. It is to these two recent contributions to biopolitics that we will now turn.

Existence Without Life

Why does a life-affirming politics always risk being transformed into a life-negating politics? Although it is only in one of his most recent books that Esposito attempts a solution to the paradox contained in Foucault's analysis of biopolitics, in reality this solution builds on a body of work that he has been developing over at least the last two decades:

> Either life holds politics back, pinning it to its impassable natural limit, or, on the contrary, it is life that is captured and prey to a politics that strains to imprison its innovative potential. Between the two possibilities there is a breach in signification, a blind spot that risks dragging the entire category into a vacuum of sense. It is as if biopolitics is missing something (an intermediary segment or a logical juncture) that is capable of unbinding the absoluteness of irreconcilable perspectives in the elaboration of a more complex paradigm that, without losing the specificity of its elements, seizes hold of the internal connections or indicates a common horizon.[49]

this paroxysmal point. But this play is in fact inscribed in the workings of all states" (*ibid.*, 260).

47 *Ibid.*, 254.

48 See *Ibid.*, 261-263.

49 Esposito, *Bíos*, 32.

The paradigm Esposito refers to in the above passage from *Bíos: Biopolitics and Philosophy* is that of *immunization*. Hence, in order to understand his intervention in the field of biopolitics in all its intricacy, we must first examine his earlier reflections on this concept.

The first point to note is that, for Esposito, the concept of *immunitas* is ineluctably tied up with that of *communitas*. In his search for a more original interpretation of the notion of community, one that goes beyond its traditional definition of what individuals have in common (*koinos*, *commun*, *gemein*), Esposito draws attention to the fact that a more thorough examination of the original Latin term from which it derives, namely *munus*, yields three additional meanings: *onus*, *officium* and *donum*. While the first two of these refer to a duty or an obligation one has to fulfill, such as taking up office (*officium*), the third meaning indicates a gift. Although these two initially seem to contradict each other – how can a gift be obligatory? – Esposito explains that it actually concerns a particular gift. In contrast to the gift as *donum*, the gift as *munus* is a gift for which one is obliged to give something in return: "once someone has accepted the *munus*, an obligation (*onus*) has been created to exchange it either in terms of goods or service."[50] In other words, in the case of the gift as *munus*, the accent falls entirely on the gift that one gives and not on what one receives. The conclusion Esposito draws from this etymological review of the concept of *communitas* is that the members of a particular community are united less by a particular quality they might share (language, descent, religion, etc.) but by the obligation to give to the other, and that therefore establishes a lack at the level of the subject: "*[C]ommunitas* is the totality of persons united not by a 'property' but precisely by an obligation or a debt; not by an 'addition' but by a 'subtraction': by a lack, a limit that is configured as an onus, or even as a defective modality for him who is 'affected.'"[51]

Thus, according to Esposito, the community is not a place where subjects encounter others endowed with similar qualities and with whom they can therefore identify, but an empty zone or void that expropriates what is most proper to them, namely their very subjective identities:

> [T]he community cannot be thought of as a body, as a corporation in which individuals are founded in a larger individual. Neither is community to be interpreted as a mutual, intersubjective "recognition" in which individuals are reflected in each other so as to confirm their initial identity […]. The community isn't a mode of being, much less a "making" of the individual subject. It isn't the subject's expansion or multiplication but its exposure to

50 Esposito, *Communitas*, 4.

51 *Ibid.*, 6.

> what interrupts the closing and turns it inside out: a dizziness, a syncope, a spasm in the continuity of the subject.[52]

Conceived this way, the community is not to be understood as a dimension in which one feels safe and secure. On the contrary, exposing oneself to the community is one of the riskiest actions one can perform. Since it involves the painful loss of borders which are supposed to ensure the subject's identity, it always involves the imminent risk of violent conflict. Because if *communitas* radically de-subjectifies and de-centers the self, if it renders it impossible for the subject to distinguish between what is 'mine' and what is 'thine,' then a violent need to restore the lost integrity of these borders will also always emerge. Esposito is therefore led to introduce into his analysis a second element of the constitutive nature of community. Essentially, *communitas* always appears to have already been traversed by an opposing force which seeks to restrain its radical expropriating tendency: *immunitas*. Etymologically speaking, *immunitas* is a privative term whose meaning derives from what it negates, namely the *munus* as an obligatory gift:

> Those who are immune owe nothing to anyone, in terms of both *vacatio* and *excusatio*: whether referring to an originary autonomy or the later release from a previously contracted debt, what counts in defining the concept is exemption from the obligation of the *munus*, be it personal, fiscal, or civil.[53]

However, in addition to this privative meaning of exemption, immunity can also, and perhaps even more primarily, be understood in a comparative sense as a privilege one enjoys. Immune, then, is he who is released from the burden of exposing himself to the communal circuit of gift-giving, and who is therefore protected from the expropriating effects of the community. However, this does not mean that Esposito views immunization as an exclusively negative or harmful process that always threatens to lead to the disintegration of the community. A certain degree of immunization cannot be avoided, for without it the community would eventually implode under the weight of an unbearable excess of expropriation: "To survive, the community, every community, is forced to introject the negative modality of its opposite, even if the opposite remains precisely a lacking and contrastive mode of being of the community itself."[54]

Yet even this does not completely exhaust the meaning of immunity. After all, in contemporary usage the term more commonly refers to the

52 *Ibid.*, 7.

53 Esposito, *Immunitas*, 5.

54 Esposito, *Bíos*, 52.

biological condition of an organism's resistance to infection or disease. What is particularly interesting about this form of immunity, and which will allow Esposito to give it an explicit juridico-political significance, is that immunity can be acquired not only naturally but also artificially, for example, through vaccination. The idea behind this second type of biological immunity is that an organism can be protected from an infection by injecting it with an attenuated dose of the same infectious disease, which means that the immune mechanism functions precisely by exposing itself to what it seeks to destroy. Put differently, immunity involves life preserving itself paradoxically by opening itself up to the very power that negates it:

> For life to remain as such, it must submit itself to an alien force that, if not entirely hostile, at least inhibits its development (...). This is where the structurally aporetic character of the immunitary process is to be located: unable to directly achieve its objective, it is forced to pursue it from the inside out. In so doing, it retains its objective in the horizon of meaning of its opposite: it can prolong life, but only by continuously giving it a taste of death.[55]

Although the notion of immunity originated in a juridico-political context, it is certainly no coincidence that medicine defines it in these terms. In fact, according to Esposito, there is a structural analogy between biological and political immunization, so much so that it is possible to see politics as an immune system that protects the members of a community by suppressing the violent conflicts engendered by an excess of communal expropriation.

Esposito claims that this understanding of politics allows us to see more clearly what is wrong with Foucault's notion of biopolitics. In Foucault's view, there is a clear historical and conceptual distinction between sovereign power and biopower. As mentioned above, the former is a negative power which suppresses life in order to reproduce itself, while the latter is a positive power which improves life in order to extend itself. But, as we have seen, by insisting so strongly on the irreducible difference between the two, Foucault was unable to explain how it is possible that a biopolitical regime could also turn against life. What he therefore failed to articulate is the idea that both of these power regimes actually function according to an immunitary logic:

> In relation to the analysis initiated by Foucault, sovereignty is understood not as a necessary compensatory ideology vis-à-vis the intrusiveness of control *dispositifs* nor as a phantasmal replica of the ancient power of death to the

55 Esposito, *Immunitas*, 8-9.

> new biopolitical regime, but as the first and most influential that the biopolitical regime assumes.[56]

According to Esposito, politics has always been biopolitics, in the sense that politics has always been preoccupied with the preservation of life and thus with the need for immunization; however, it will only become explicitly so at the moment this issue is no longer seen as something given but as a problem to be addressed and solved:

> By this it is understood that all civilizations past and present faced (and in some way solved) the needs of their own immunization, but that it is only in the modern ones that immunization constitutes its most intimate essence. One might come to affirm that it wasn't modernity that raised the question of the self-preservation of life, but self-preservation is itself raised in modernity's own being, which is to say it invents modernity as a historical and categorical apparatus able to cope with it.[57]

It should therefore not be too much of a surprise that Esposito sees Thomas Hobbes as the founding father of modern immunitary biopolitics.

It is well known that Hobbes places the problem of self-preservation (*conservatio vitae*) at the center of his political philosophy. In fact, according to Hobbes, it constitutes man's most essential natural right. Man is free to use any means necessary to preserve himself in being. However, in the state of nature this attempt at self-preservation is bound to fail because it is accompanied by an equally strong natural impulse that tends to contradict the first one: the inexhaustible desire for acquisition: if all individuals were to pursue only their own self-interest and simply take by force whatever was deemed necessary for their own survival, there would inevitably be continuous violent conflict. In other words, this would lead to a "war of all against all" which would cause man's life to be "solitary, poor, nasty, brutish, and short." It is in order to counter this fatal logic that Hobbes develops an immunitary strategy. If left to its own devices, life is destined to contradict its own most natural impulse. That is why it is necessary for life to renounce its natural right to unlimited acquisition and to transfer the task of self-preservation to an external power: "[I]n order to save itself, life needs to step out from itself and constitute a transcendental point from which it receives orders and shelter."[58] Faced with the imminent threat of generalized conflict, humans in their natural state are forced to institute a sovereign power which acts to protect them from a possible return to conflict. According to Esposito, Hobbes's theory

56 Esposito, *Bíos*, 57.

57 *Ibid.*, 54-55.

58 *Ibid.*, 58.

of the social contract is therefore little more than a thinly veiled attempt to immunize the community from the danger of communal expropriation. In describing the threat to humans in their natural state as a consequence of the conflict inherent to their excessive desire to appropriate one another's goods, Hobbes ultimately proposes a model in which the horizontal yet conflict-ridden relationships among equals are replaced by the vertical relationships between a sovereign and its individual subjects. In this sense, one might even say that "sovereignty, in the final analysis, is nothing other than the artificial vacuum created around every individual – the negative of the relation or the negative relation that exists between unrelated entities."[59]

In fact, though, this is only the first stage of Hobbes's immunitary strategy – its 'positive' side, so to speak. As we know from medical practice, immunization involves injecting the body with a sample of the very same infection we wish to protect it from and thus submitting the body to a hostile force that, in a way, subtly counters the infection from the inside out. In order to protect life, the immune system needs to reproduce the malign force that besieges it in an attenuated yet essentially no less lethal form. The same holds true for the immunization of the political body. If sovereign power is created to protect the community from violent conflict, it does not mean that it can provide for the complete elimination of violence in the commonwealth. Rather, violence is reproduced in the person of the sovereign himself, who is the only individual in the commonwealth to retain the natural right to kill. The paradox is that the subjects do not even have the right to resist this sovereign power over life and death because they have granted the sovereign this power of their own free will:

> This exception – the liminal coincidence of preservation and capacity to be sacrificed of life – represents both a remainder that cannot be mediated and the structural antinomy on which the machine of immunitary mediation rests. [...] It is as if the negative, keeping to its immunitary function of protecting life, suddenly moves outside the frame and on its reentry strikes life with uncontrollable violence.[60]

Contrary to what Foucault thought, then, biopower and sovereign power are not mutually exclusive but should be seen as two sides of the same coin. If a politics that sets itself the task of preserving life always ends by negating life, it is because it can only do so by subjecting life to the very threat that it sought to overcome.

59 *Ibid.*, 61.

60 *Ibid.*, 62-63.

However, this does not entirely answer the question of how we should understand the specificity of Nazi biopolitics. As we recall, Foucault always oscillated between the continuist and discontinuist approaches without ever decisively opting for either one. In order to save his theory from becoming invalidated, he turned to the concept of racism to explain the emergence of totalitarianism. But not only does this neglect the question of how we should conceive of a communist racism, it also threatens to erase the difference between totalitarian biopolitics and modern biopolitics in general, which, in *Society Must Be Defended*, Foucault also traced back to a racist matrix. Relating both sovereign power and biopower to the paradigm of immunization, Esposito thus has no need of a mediating concept such as racism in explaining the dialectical reversal from a life-affirming politics to a life-negating politics. Yet this still fails to explain why only Nazism has propelled the thanatological drive of biopolitics to its most complete realization. For this is what Nazi biopolitics gives us to think: it not only sought to cleanse the German national body of all internal and external enemies but – as we know from Telegram Number 71 sent from Hitler's bunker – it also tended toward its own self-destruction.

Esposito detects both quantitative and qualitative differences between modern biopolitics and Nazi biopolitics. Firstly, no other biopolitical regime has ever ventured to push the use of biological metaphors further than Nazism did. Biological terms such as "body" and "constitution" have frequently been used to illustrate some particular political concept, but only Nazism took these metaphors so seriously that its leaders "use[d] biological processes as criteria with which to guide their own actions."[61] In this sense, it could be argued that Nazism did not politicize biology but in fact biologized politics. This also accounts for Nazism's essential difference from communist totalitarianism: "While the transcendental of communism is history, its subject class, and its lexicon economic, Nazism's transcendental is life, its subject race, and its lexicon biological."[62] This is already clear from the fact that the German medical class was accorded a key role in the regime's genocidal practices. Esposito notes, for example, that the Reich's legal code required the active participation of a physician in all stages of the process. Only doctors were allowed to decide who was to be given a 'merciful' death in the T4 euthanasia program and who, upon entry to the concentration camp, was to be sent immediately to the gas chambers; in addition, though, they were also the only individuals qualified to administer a lethal injection or to open a gas valve. From this point of view, it could even be said that Nazi doctors were sincerely concerned

---

[61] *Ibid.*, 112-113.

[62] *Ibid.*, 112.

about the health of their patients, if by this it is understood that they identified the patient to be the German *Volkskörper*, and that they were convinced that restoring this body to health required the elimination of all of the harmful elements that were devouring this body from the inside out. If this still gives the impression that Nazism was merely a more radicalized version of modern immunitary biopolitics, however, this is only partially correct according to Esposito. For if it is true that the Nazi immunitary apparatus above all "propagated the fight to the death against the Jews as the resistance put up by the body (and originally the healthy blood) of the German nation against the invading germs that had penetrated within and whose intent it was to undermine the unity and life of the German nation itself,"[63] even to the point of destroying its own body, then it would be more correct to say that Nazism is characterized by an *autoimmune* logic.

Esposito captures the logic that underpinned the Nazi (auto-)immunitary apparatus in a single formulation: "*regeneration* overcomes *degeneration* through *genocide*."[64] As indicated in an earlier section in this chapter, the notion of 'degeneration' played a key role in the self-legitimation of the Nazi regime. Originally coined in late 18th century biology, the term featured prominently in the philosophical, juridical and medical discourses of prewar Germany. The essential idea was that a growing segment of the population was affected by hereditary malformations which could eventually lead to sterility and even to the extinction of the entire hereditary line. Above all, a degenerate was defined as someone who deviates from the norm. This was not to say that their aberrant physical or psychological traits might represent natural variation within a population that could lead to successful adaption: rather, they were considered to be suffering from atavism, in the sense that they were situated within a zone where the boundary between the human and the animal was indistinct:

> Degeneration is the animal element that reemerges in man in the form of an existence that isn't properly animal or human, but exactly their point of intersection: the contradictory copresence between two genera, two times, two organisms that are incapable of producing a unity of the person and consequently for the same reason incapable of forming a juridical subjectivity.[65]

This largely explains the extraordinary development of German anthropology and zoology of that time. If doubts arose about the stability of the borders separating humans from animals, efforts had to be made to reestablish them. This not only yielded a new taxonomy of human

63 *Ibid.*, 116.

64 *Ibid.*, 137; Emphasis added.

65 *Ibid.*, 119.

biotypes – which ranged from the Aryan overman and the Mediterranean average man to the Slavic subhuman and the Jewish anti-man – but also led to the incorporation of animal and vegetable species within the human race. Contrary to what has generally been assumed, then, Nazi science did not animalize man but actually widened the definition of man to include animals of inferior species: "He who was the object of persecution and extreme violence wasn't simply an animal (which indeed was respected and protected as such by one of the most advanced pieces of legislation of the entire world), but was an *animal-man*: man in the animal and the animal in man."[66] Moreover, not only was it believed that deficient traits and behaviors were inexorably transmitted from generation to generation, these deficiencies were also considered to have the potential to spread from one body to the next until they affected the entire population. It was therefore thought that unless measures were taken to stop this process from running its full course, civilization would eventually come to an end.

If degeneration was the infectious disease that harmed the national body, this body could only be restored to health through a regenerative program that aimed to remove the sick parts. This was to be achieved through eugenics or, what amounted to the same thing, racial hygiene: "Racial hygiene is the immunitary therapy that aims at preventing or extirpating the pathological agents that jeopardize the biological quality of future generations."[67] In contrast to its American counterpart, which focused almost exclusively on measures for improving the hereditary quality of the population, the Nazi eugenics program mainly aimed to prevent the degenerative process from causing further damage to the national body. Or, in other words, it was believed that the most effective way to produce the healthiest beings required the elimination of the unhealthiest. It was therefore no coincidence that the law on forced sterilization was one of the first legislative measures taken by the Nazi's when they came into power because, since it "intervene[d] at the root, at the originary point in which life is spread, [...] it concerned the principle according to which the political body had to be vaccinated beforehand from every disease that could alter the self-preserving function."[68] If we recall that Fritz Lenz, then the head of the Kaiser Wilhelm Institute of Anthropology, Human Heredity and Eugenics, had estimated that approximately one third of the German population would have to be sterilized in order to achieve the aim of complete regeneration, then we can begin to understand how important

---

66 *Ibid.*, 130.

67 *Ibid.*, 128.

68 *Ibid.*, 132.

the phenomenon of birth was for the Nazi regime – indeed, so important that none of its aspects escaped state control and supervision.

If sterilization constituted one pole of the negative eugenic apparatus, euthanasia constituted the other. With regard to this issue, Esposito stresses the importance of two books which appeared only a couple of decades before the T4 euthanasia program was officially launched. The first of these is Adolf Jost's *Das Recht auf den Tod*[69] [The Right to Die]. Jost was not only the first to introduce the concept of a "life without value" (*negativen Lebenswert*) but also one of the first scholars to argue that the right to end life belonged exclusively to the state. Even more significant, however, is Karl Binding and Alfred Hoche's *Die Freigabe der Vernichtung lebensunwerten Lebens*[70] [Authorization for the Annihilation of Life Unworthy of Being Lived], which contains a detailed legal discussion of the question of whether it might be desirable to extend the law on the legality of suicide to the killing of the incurably ill and the mentally disabled. While the jurist Binding quickly becomes engaged with the complex technical question of whether the patient's consent would be necessary to terminate his or her life, and, if so, how consent could be obtained in cases where the patient was incapable of giving it, Hoche's analysis takes a somewhat different approach to the issue by arguing that the lawful killing of such persons could be justified on biological grounds alone. The terms he uses to describe their condition of existence – "half-men," "damaged beings," "mentally dead," "human ballast," "empty human husks" – are meant to indicate that it would not really mean killing a human being because "the persons whom it strikes are *already* dead."[71] There is hardly any doubt that this is exactly what the regime had in mind when it issued an instructional film for personnel attached to the T4 program, the telling title of which was *Dasein ohne Leben* [Existence without Life]. Such an existence without life, a form of life entirely lacking in 'spirit,' was nothing more than bare biological existence and thus could be disposed of as "a life unworthy of being lived." Of course, spirit should not be understood here in any dualistic sense as a faculty that opens the biological body up to transcendence. As we know from Alfred Rosenberg's *The Myth of the Twentieth Century*, for example, it is only by belonging to a racial hereditary line that one's body receives spirit. It therefore amounts to much the same thing whether we say that what the regime wished to eliminate was existence without life or existence

69 Adolf Jost, *Das Recht auf den Tod* (Göttingen: Grunow & Co., 1895).

70 Karl Binding and Alfred Hoche, *Die Freigabe der Vernichtung Lebensunwerten Lebens: Ihr Mass und ihre Form* (Leipzig: Meiner, 1920).

71 *Ibid.*, 134.

without spirit. What matters from this point of view is that death was both the target and the instrument of the regime's regenerative therapy – the sickness and its remedy:

> The disease against which the Nazis fight to the death is none other than death itself. What they want to kill in the Jew and in all human types like them isn't life, but the presence in life of death: a life that is already dead because it is marked hereditarily by an original and irremediable deformation; the contagion of the German people by a part of life inhabited and oppressed by death. [...] [T]he Nazi regime strengthened its own immunitary apparatus to the point of remaining victim to it. The only way for an individual or collective organism to save itself definitely from the risk of death is to die. It was what Hitler asked the German people to do before he committed suicide.[72]

Fighting death by killing in ever greater degrees, the Nazi immunitary apparatus finally became autoimmunitary and turned against itself.

Sacred Life

Giorgio Agamben is perhaps the single most significant contemporary thinker to emerge in the wake of Foucault's concept of biopolitics. Although he was already fairly well known for his work in aesthetics and the philosophy of language,[73] it was only after the publication of the first volume of his multi-volume project *Homo Sacer* that Agamben would become truly renowned in philosophical circles and beyond,[74] not least because of its extremely provocative undertone. Agamben's claims that there is "an inner solidarity between democracy and totalitarianism"[75] and that the concentration camp constitutes "the fundamental biopolitical paradigm of the West,"[76] have been fiercely criticized for their crude and reductionist view of the modern political condition.[77] Although we will not discuss here whether these

---

72 *Ibid.*, 137-138.

73 See especially Giorgio Agamben, *Language and Death: The Place of Negativity*, trans. K. E. Pinkus and M. Hardt (Minneapolis: The University of Minnesota Press, 1991) and Giorgio Agamben, *Idea of Prose*, trans. M. Sullivan and S. Whitsitt (Albany: SUNY Press, 1995).

74 Giorgio Agamben, *Homo Sacer: Sovereign Power and Bare Life*, trans. D. Heller-Roazen (Stanford: Stanford University press, 1998).

75 Agamben, *Homo Sacer*, 10.

76 *Ibid.*, 181.

77 See for example Jacques Rancière (2004), 'Who is the subject of human rights?' *South Atlantic Quarterly* 103: 297-310 and Niels Werber (2002), 'Die Normalisierung des Ausnahmefalls: Giorgio Agamben sieht immer und überall Konzentrationslager,' *Merkur* 56: 618-622.

particular critiques really invalidate the basic tenets of his argument, it is already clear that they touch upon the core issue of our discussion so far: how is it possible that biopolitics can divide into both a life-affirming politics and a life-denying politics? We have shown that Foucault was unable to explain the emergence of totalitarian thanatopolitics without invalidating his fundamental thesis that, in the modern period, the sovereign power to take life was replaced by a biopower that fosters life. Esposito sought to resolve this paradox by arguing that the paradigm of immunity constitutes the ultimate terrain within which to understand both sovereign power and biopower as immunitary *dispositifs* centered on the specifically modern problem of the preservation of life. Describing Nazism by means of an autoimmunitary logic, Esposito makes it clear that Nazism shares much with modern biopolitics but simultaneously denies that one can be wholly reduced to the other. Agamben also locates the "blind spot" of Foucault's theory in "the hidden point of intersection between the juridico-institutional and the biopolitical models of power"[78] and therefore presents his thesis on biopolitics as a "correction or, at least, completion"[79] of the Foucauldian thesis. But, in contrast to Esposito, he believes that this convergence of sovereign power and biopower is not an exclusively modern phenomenon. Arguing that "the production of a biopolitical body is the original activity of sovereign power"[80] and that "biopolitics is at least as old as the sovereign exception,"[81] Agamben traces the origin of this mechanism back to the very beginning of the Western political tradition.

The starting point of Agamben's critical analysis of biopolitics is what he calls "the paradox of sovereignty," which consists in the fact that the sovereign is both inside and outside of the legal order simultaneously. Agamben relies heavily on the work of the German legal scholar Carl Schmitt to explain what is at stake in this paradox. Writing in the context of latent civil war characteristic of early Weimar Germany, Schmitt argued that the liberals' ignorance of the possibility that some exceptionally threatening event might occur kept them from developing a consistent theory of emergency powers that could protect the legal order in times of extreme peril. More specifically, Schmitt detects two

78 Agamben, *Homo Sacer*, 6.

79 *Ibid.*, 9.

80 *Ibid.*, 6.

81 *Ibid.*

fundamental flaws in the liberal theory of the state.[82] Firstly, some of its most essential principles, such as the separation of powers and the system of checks and balances, impede the state in deciding clearly *who* has the power to proclaim a state of exception and take the necessary measures to restore law and order. Secondly, since the exception cannot be subsumed under a preexisting legal norm, it is impossible to appeal to codified law to determine in advance *what* must done to suppress extremely dangerous threats to the legal order. The first line of Schmitt's *Political Theology*, "[s]overeign is he who decides on the exception,"[83] should be read as a remedy to both of these flaws in the liberal constitutional state. By reintroducing a concept of sovereignty that explicitly assumes the existence of a powerful sovereign ruler who can proclaim the state of exception and, if necessary, suspend the law, Schmitt claimed to have found a viable alternative to liberalism's bureaucratic rule-bound formalism. As Agamben clarified in *The State of Exception*, what is ultimately at issue here is "the question of the juridical significance of a sphere of action that is in itself extrajuridical."[84] For Agamben, then, sovereignty is truly a borderline concept, for "the sovereign stands outside the juridical order and, nevertheless, belongs to it, since it is up to him to decide if the constitution is to be suspended *in toto*."[85] The paradox is that, in order to protect the legal order, the sovereign has the power to legally place himself outside of it.

However, what especially interests Agamben about the sovereign exception is that it creates "the very space in which the juridico-political order can have validity."[86] In order for the law to be applicable, a normal order must be established because no law is applicable to chaos alone. By deciding on an exception, the sovereign excludes an individual case from the law. Being beyond the reach of law, however, does not mean that the exception is entirely without relation to it. On the contrary, in the state of exception, the individual case which is excluded from the law remains tied to the law precisely in the form of the law's suspension.

---

82 Both critiques are captured in the following quote: "The essence of liberalism is negotiation, a cautious half measure, in the hope that the definitive dispute, the decisive bloody battle, can be transformed into a parliamentary battle and permit the decision to be suspended forever in an everlasting discussion." See Carl Schmitt, *Political Theology. Four Chapters on the Concept of Sovereignty*, trans. G. Schwab (Chicago: The University of Chicago Press, 2005), 63.

83 *Ibid.*, 5.

84 Giorgio Agamben, *The State of Exception*, trans. K. Attell (Chicago: The University of Chicago Press, 2005), 11.

85 Schmitt, *Political Theology*, 7.

86 Agamben, *Homo Sacer*, 19.

As Agamben puts it, "*[t]he rule applies to the exception in no longer applying, in withdrawing from it.*"[87] According to Agamben, this means that the state of exception is not the chaos preceding the legal order but rather the result of the sovereign decision to suspend the law in its entirety. Or, in other words, by suspending itself, the rule gives rise to the exception and, "maintaining itself in relation to the exception, first constitutes itself as a rule."[88] Ultimately, then, the function of the sovereign decision is to determine what is included in the juridical order and what is excluded from it. But it can only do so by creating an indistinct zone between order and chaos – the state of exception – on the basis of which the distinction between what is included in and excluded from the juridical order first becomes meaningful. The situation of the exception is therefore one of neither fact nor right, but constitutes a threshold between the two.

Agamben argues that this sovereign legal structure can best be understood by comparing it to the structure of language. In the same way that a word can only be used to refer to a particular aspect of reality because the word is also meaningful independently of that aspect, a juridical rule can only be applied to a concrete case because it is also valid independently of that particular case. In both instances, the abstract principle can only refer to a concrete situation because it is already in force as pure potentiality in relation to every actual reference:

> [J]ust as language presupposes the nonlinguistic as that with which it must maintain itself in a virtual relation (in the form of *langue* or, more precisely, a grammatical game, that is, in the form of a discourse whose actual denotation is maintained in infinite suspension) so that it may later denote it in actual speech, so the law presupposes the nonjuridical (for example, mere violence in the form of the state of nature) as that with which it maintains itself in a potential relation in the state of exception.[89]

What Agamben means here is that, for example, in order for a law prohibiting homicide to become applicable, the concrete fact to which it refers (the act of killing) cannot be qualified as homicide as such, because it only becomes homicide once it is subsumed under the law prohibiting it. This means that, in taking exception to itself, the law prohibiting homicide presupposes the act of an unsanctionable killing as a condition of its applicability in normal conditions:

87 *Ibid.*, 18.

88 *Ibid.*

89 *Ibid.*, 20-21.

> *The sovereign exception (as zone of indistinction between nature and right) is the presupposition of the juridical reference in the form of its suspension.* Inscribed as a presupposed exception in every rule that forbids something (for example, in the rule that forbids suicide) is the pure and unsanctionable figure of the offence that, in the normal case, brings about the rule's own transgression (in the same example, the killing of a man not as natural violence but as sovereign violence in the state of exception).[90]

The term Agamben uses to describe the law's capacity to apply in no longer applying is the "ban." That which is excluded from the law in the sovereign exception is not simply placed outside of the law and made irrelevant to it, but is in fact abandoned by it, in the sense that it becomes impossible to say whether it sojourns inside or outside the juridical order.

Although Agamben's conception of the sovereign exception is much indebted to Schmitt's theory of sovereignty, there is also a crucial difference. While for Schmitt the exception reveals a lacuna in the liberal theory of the state, Agamben sees it as constituting an essential condition for the law's application. That is, for Agamben, the lacuna resides less – or not only – in the relation between liberal positive law and the state of emergency but instead, more fundamentally, in that between law and reality. It is because the law cannot immediately be applied to reality that "it is ultimately necessary to suspend its application, to produce an exception."[91] However, in speaking of "reality" it remains much too vague, for the gap that the sovereign exception attempts to fill is more precisely that between law and life: "Law is made of nothing but what it manages to capture inside itself through the inclusive exclusion of the *exceptio*: it nourishes itself on this exception and is dead letter without it. In this sense, the law truly 'has no existence in itself, but rather has its being in the very life of men'."[92] But if life can only be brought into the sphere of law by means of the sovereign decision on the exception, then this operation will always produce a "limit-figure of life, a threshold in which life is both inside and outside the juridical order."[93] In a sense, then, for the paradoxical figure of the sovereign who is both inside and outside of the legal order, there is an equally paradoxical corresponding figure of life which is also both inside and outside of the legal order. Central to Agamben's understanding of this perfect coincidence of law and life in the state of exception is a discussion between Gershom Scholem and Walter Benjamin about the structure of the law in Kafka's writings. In a

90 *Ibid.*, 21.

91 Agamben, *State of Exception*, 40.

92 Agamben, *Homo Sacer*, 27.

93 *Ibid.*

letter to Benjamin, Scholem writes that the impenetrability of the law in Kafka's *The Trial* can best be described as "the Nothing of Revelation," by which he means that the law is emptied of all significance yet still remains in force. According to Agamben, Scholem's phrase "being in force without significance"[94] (*Geltung ohne Bedeutung*) perfectly captures the structure of the law in the state of exception. In being reduced to its pure form, the law is not absent but in fact suspended in its application, and thus only "appears in the form of its unrealizability."[95] But, as Benjamin responds, if the law has lost its content and intelligibility, and only retains its form, then "the law has become indistinguishable from life."[96] A law in force without significance corresponds to a life entirely transformed into law.

But how should we understand this "limit-figure of life" which, according to Agamben, founds the law? At the beginning of *Homo Sacer*, Agamben notes that the Greeks had two distinct terms for what we call "life:" *zoē*, "which expressed the simple fact of living common to all living beings (animals, men, or gods)" and *bios*, "which indicated the form or way of living proper to an individual or a group."[97] As we know from Aristotle's political treatises, the Greeks saw *zoē* or natural life entirely as an affair of the private household (*oikos*), while *bios* was the form of life that concerned the *polis*. It is therefore interesting that when, in describing the process that led to the emergence of biopolitics, Foucault notes that "for millennia, man remained what he was for Aristotle: a living animal with the additional capacity for a political existence; modern man is an animal whose politics places his existence as a living being in question."[98] Here, he means that biopolitics consists precisely in an overturning of this fundamental Greek distinction between *zoē* and *bios*. While ancient politics insisted on the radical separation of biological life and political life, modern politics sees in the former its principal object. In Agamben's words, "if anything characterizes modern democracy as opposed to classical democracy, then, it is that modern democracy presents itself from the beginning as a vindication and liberation of *zoē*, and that it is constantly trying to transform its own bare life into a way of life and to find, so to speak, the *bios* of *zoē*."[99] Modern biopolitics thus seeks to include in the political sphere the very element – *zoē* or natural life – whose exclusion

94 *Ibid.*, 51.

95 *Ibid.*

96 *Ibid.*, 53.

97 *Ibid.*, 1.

98 Foucault, *The History of Sexuality*, 143.

99 Agamben, *Homo Sacer*, 9.

from it was the original founding act of the sovereign space of the *polis*. According to Agamben, this political project clearly rests on a paradox, for "it wants to put the freedom and happiness of men into play in the very place – 'bare life' – that marked their subjection."[100] Bare life, then, the product of this operation, is neither *zoē* nor *bios* but "the zone of indistinction in which *zoē* and *bios* constitute each other in including and excluding each other."[101]

Importantly, Agamben argues, in contrast to Foucault, that the politicization of natural life and the ensuing production of bare life is not exclusive to the modern period but in fact "the hidden foundation on which the entire political system rested."[102] In order to demonstrate this, he turns to the obscure figure of *homo sacer* from archaic Roman law. Quoting Pompeius Festus, Agamben explains that the "sacred man"

> is the one whom the people have judged on account of a crime. It is not permitted to sacrifice this man, yet he who kills him will not be condemned for homicide; in the first tribunitian law, in fact, it is noted that 'if someone kills the one who is sacred according to the plebiscite, it will not be considered homicide.' This is why it is customary for a bad and impure man to be called sacred.[103]

On the one hand, then, the *homo sacer* is excluded from human law, for the one who kills him will not be considered a murderer; on the other hand, however, he is also excluded from divine law, for his killing could not be viewed as a form of ritual purification. In other words, the *homo sacer* was someone who was banned from human law without being allowed to enter into the sphere of divine law. But, Agamben adds, not only is the *homo sacer* excluded from both human and divine law, he is simultaneously included in these realms: "*homo sacer* belongs to God in the form of unsacrificeability and is included in the community in the form of being able to be killed."[104] Doubly captured and doubly excluded, the *homo sacer* is thrown into a liminal zone between profane and divine law where he finds himself abandoned to unsanctionable violence. According to Agamben, this shows that there is a striking symmetry between the figure of *homo sacer* and that of the sovereign, in the sense that both are simultaneously included and excluded from the law: "The sovereign sphere is the sphere in which it is permitted to kill without committing

100 *Ibid.*, 10.

101 *Ibid.*, 90.

102 *Ibid.*, 9.

103 Quoted in *Ibid.*, 71.

104 *Ibid.*, 82.

homicide and without celebrating a sacrifice, and sacred life – that is, life that may be killed but not sacrificed – is the life that has been captured in this sphere."[105]

Even if Agamben sees the sovereign capture of bare life as the "originary exclusion through which the political dimension was first constituted,"[106] and thus holds that sovereignty and biopolitics were originally intertwined, this does not mean that he recognizes no difference between the ancient and modern versions of this biosovereign mechanism. The crucial difference is that while in the ancient world the sovereign production of bare life was only an exceptional event, in the modern world it has increasingly become the rule:

> [W]hat characterizes modern politics is not so much the inclusion of *zoē* in the *polis* – which is, in itself, absolutely ancient – nor simply the fact that life as such becomes a principal object of the projections and calculations of State power. Instead the decisive fact is that, together with the process by which the exception everywhere becomes the rule, the realm of bare life – which is originally situated at the margins of the political order – gradually begins to coincide with the political realm, and exclusion and inclusion, outside and inside, *bios* and *zoē*, right and fact, enter into a zone of irreducible indistinction.[107]

At the center of Agamben's interpretation of modern biopolitics is the provocative thesis that bare life has become the central political subject, which implies that man's inclusion in the modern political order could only be achieved by his being constantly exposed to sovereign violence. Agamben traces the first appearance of the modern *homo sacer* back to the famous writ of *habeas corpus* in 17th century England. This principle, which holds that a person under arrest must be brought before a court, is generally considered to be one of the most fundamental tenets of the Western legal tradition in that it is intended to protect the individual from arbitrary state action. Agamben, however, points out the biopolitical operation at work in this principle ("you will have to have a body to show"), which makes the body the central object of the juridico-political: "he who will appear later as the bearer of rights […] can only be constituted as such through the repetition of the sovereign exception and the isolation of *corpus*, bare life, in himself."[108] Even more radically, Agamben claims that this is also what is ultimately at stake in the French Declaration of the Rights of Man and Citizen. While this is usually regarded as the founding

105 *Ibid.*, 83.

106 *Ibid.*

107 *Ibid.*, 9.

108 *Ibid.*, 124.

document of modern democracy and human rights, Agamben takes the notion that "men are born and remain free and equal in rights" to imply that it is precisely bare life – the mere fact of being born – that is given rights and made the principal political subject. Moreover, by stating that "the principle of all sovereignty resides essentially in the nation," the declaration makes these "inalienable rights" immediately contingent upon one's belonging to the nation (from *nascere*, to be born): "the fiction implicit here is that *birth* immediately becomes *nation* such that there can be no interval of separation between the two terms."[109] This is what constitutes the hidden *aporia* of modern politics: since its project of politicizing natural life is ultimately grounded on a fiction, it will continuously produce bare life as an inevitable side effect of its failure to establish this connection. According to Agamben, the figure of the refugee represents the most clear example of this failure. Stripped of his rights as a citizen of the nation-state and thereby bringing to light the fictitious connection between nativity and nationality, the refugee "causes the secret presupposition of the politics domain – bare life – to appear for an instant within that domain."[110]

Clearly, then, like Foucault and Esposito, Agamben views biopolitics as an irreducibly double-sided phenomenon:

> [T]he spaces, the liberties, and the rights won by individuals in their conflicts with central powers always simultaneously prepared a tacit but increasing inscription of individuals' lives within the state order, thus offering a new and more dreadful foundation for the very sovereign power from which they wanted to liberate themselves. [...] The fact is that one and the same affirmation of bare life leads, in bourgeois democracy, to a primacy of the private over the public and of individual liberties over collective obligations and yet becomes, in totalitarian states, the decisive political criterion and the exemplary realm of sovereign decisions.[111]

If Agamben detects "an inner solidarity between democracy and totalitarianism,"[112] then it is because both types of regime are founded on the same living substance – bare life. Yet this does not necessarily mean that they can be wholly reduced to one another. Rather, Agamben's point is that totalitarian rule is quantitatively but not qualitatively different from democratic rule of law. According to him, the former simply pushes to their logical conclusion some of the tendencies that can also be found

[109] *Ibid.*, 128.

[110] *Ibid.*, 131.

[111] *Ibid.*, 121-122.

[112] *Ibid.*, 10.

in the latter. By this he means that totalitarian rule was not 'lawless;' on the contrary, it sought to extend the law to the point of becoming indistinguishable from life. Only if we recognize that what Nazism aimed at was a total normativization of life can we begin to understand Agamben's most provocative thesis that the concentration camp was not a historical anomaly but "the hidden matrix and *nomos* of the political space in which we are still living."[113] It is commonly thought that the Nazi project of total domination in the concentration camps represented an – unexplainable – exception to the rule. For Agamben, however, "the camp is the space that is opened when the state of exception begins to become the rule."[114]

To understand why this is so, Agamben invites us not to study the horrible fate of those who suffered in the camps – this he will do in another volume of *Homo Sacer*[115] – but to examine instead the juridico-political structure of the camp itself. The first thing to note is that the juridical basis for internment in the camp was *Schutzhaft* or "protective custody." This exceptional measure, which has its origin in 19th century Prussian law, allowed the police to detain people solely on the suspicion that they posed a threat to the security of the state. Importantly, this measure could only be applied after the government had proclaimed the state of exception and thus after it had first suspended the articles in the constitution guaranteeing personal liberties. It is worth noting, too, that the Nazi regime did not invent this juridical construction. Given that during its short period of existence the liberal Weimar republic proclaimed the state of exception many times, it could be said that by incorporating this construction in its famous 'Decree for the Protection of the People and State' the Nazi regime merely continued a practice that was already well established in many parliamentary democracies of that time. The crucial difference, however, was that the decree permanently suspended these articles of the constitution without mentioning the notion of the state of exception as a precondition, so that in the Nazi regime the state of exception became "confused with the juridical rule itself."[116] But if we accept that the camps nonetheless fall within the scope of *Schutzhaft*, then the further implication of this is that "in the camp, the state of exception, which was essentially a temporary suspension of the rule of law on the basis of a factual state of danger, is now given a permanent spatial arrangement, which

113 *Ibid.*, 166.

114 *Ibid.*, 168-169.

115 See Giorgio Agamben, *Remnants of Auschwitz: The Witness and the Archive*, trans. D. Heller-Roazen (New York: Zone Books, 2002).

116 Agamben, *Homo Sacer*, 168.

as such nevertheless remains outside the normal order."[117] The camp is therefore nothing less than the space where the state of exception is realized normally and thus where law and fact, norm and exception, enter into a zone of indistinction. Moreover, those who were brought into the camps had already been denaturalized and stripped of their rights as citizens. In this way, the camp was also "the most absolute biopolitical space ever to have been realized, in which power confronted nothing but pure life, without any mediation." This explains why, in such a camp, to use Arendt's famous phrase, "everything is possible."

Form-Of-Life

At the end of the last chapter, we argued that liberal political and moral philosophy might not be the most appropriate framework in which to address the challenges associated with human genetic modification. The reason for this is that attempts to argue either for or against genetic enhancement seem to lead to highly illiberal conclusions: while the argument against it entails a preemptive dehumanization of any potential enhanced form of life, the argument for it threatens to reduce any non-enhanced form of life to the status of wrongful life. We therefore suggested that in order to understand this paradoxical outcome, it is necessary to turn to an interpretative framework that focuses specifically on the biological dimension of human beings as both objects and subjects of political action. Although, as we have seen, the growing convergence between life and politics in modernity has been a hot topic in political thought at least since the end of the 19th century, it was Foucault who first addressed this issue in a systematic and critical way. Having shown, however, that Foucault was never quite capable of explaining how the rise to dominance of a type of politics that seeks precisely to protect and strengthen life could culminate in the emergence of the most life-suppressing regime that history has ever seen, we provided a detailed account of how two of the most important contemporary biopolitical thinkers have, each in their own way, sought to resolve this paradox. If we now attempt a tentative extension of Esposito's and Agamben's theories of biopolitics in order to deal with the liberal eugenics debate, it must be that we are led to do so because this debate seems to founder on the very same paradox. That is not to say that liberalism in general, and liberal eugenics in particular, can be wholly reduced to Nazi biopolitics, as Agamben sometimes seems to suggest. Instead, as Esposito has argued, "liberalism is situated within the

[117] *Ibid.*, 169.

very biopolitical regime that, though configured in the opposite way, gave rise to Nazism."[118]

In what manner, then, are we to understand this latter claim? It is not hard to see that the discourse of the opponents of genetic enhancement is largely indebted to an immunitary lexicon that Esposito sees as central to modern biopolitics. In order to ascertain that this is indeed the case, we merely have to recall that in his argument against genetic enhancement Habermas sets out his uncompromising support for "a right to a genetic inheritance *immune* from artificial intervention."[119] In order to prevent genetic intervention creating the child's permanent and irreversible dependence on the parent and thus causing the moral community of equals to disintegrate, Habermas is led to introduce a transcending legal norm of biological life. There are two problems with Habermas's position, however. First, as Esposito has shown, the members of a community are united less "by reciprocal and symmetrical relations of mutual recognition,"[120] as Habermas assumes, and more "by an 'obligation' […] that expropriates them of their initial property (in part or completely), of the most proper property, namely, their very subjectivity."[121] The community is not constituted of an aggregate of individuals who share an essential 'property' or 'nature' but, as Esposito puts it, by those "who have nothing in common."[122] Second, and in addition, Habermas's attempt to immunize the community against the danger of disintegration by giving normative content to human nature actually risks reproducing the very danger he seeks to avoid, inasmuch as he seems to preemptively deny any future enhanced person the legal status of full membership in a moral and political community.

That is not to say that such immunitary logic is absent from the discourse of the advocates of a liberal eugenics. In the latter case, however, it is not an abstract notion of human nature that needs to be immunized against the threat of genetic intervention; instead, it is the body of the individual subject that must be protected from the dangers and risks associated with its free and unhindered 'natural development.' Furthermore, a case could be made that their position is largely driven by the same three "immunitary *dispositifs*" that Esposito has detected in Nazi biopolitics. The first of these is what Esposito calls the "absolute normativization

118 Roberto Esposito, *Community, Immunity, Biopolitics*, trans. R. N. Welch (New York: Fordham University Press, 2013), 107.

119 Habermas, *The Future of Human Nature*, 27; Emphasis added.

120 *Ibid.*, 65.

121 Esposito, *Communitas*, 6-7.

122 *Ibid.*, 28.

of life."[123] This *dispositif* could be said to underpin not only the position of those who invoke a normative conception of human nature to oppose genetic intervention, as we showed above, but also the project of a liberal eugenics. After all, as we saw above, in the same way that the German medical class was accorded a key role in all facets of the Nazi eugenic campaign against degeneration, so too will the power and influence of geneticists and molecular biologists increase in a future regime of liberal eugenics as they will be the ones to establish the limits within which life may unfold. The second Nazi immunitary *dispositif* that makes its reappearance in the context of the debate over genetic technologies is that of the "double enclosure of the body."[124] The first closure of the body that Esposito detects in the mechanisms of Nazi biopolitics relates to the establishment of an "absolute identity between our body and ourselves."[125] In the view of Nazi biopolitics, we don't 'have' our body but 'are' our body: the biological given represents the ultimate truth from which we can never escape. This first closure of the body is then reinforced by a second one, by which Esposito means that, in Nazi biopolitics, the self was not only incorporated in its own body, but also "into a larger body that constitutes the organic totality of the German people."[126] To illustrate this second operation, Esposito refers to the fact that, during Nazism, posters were issued which stated: "Dein Körper gehört dem Führer" (Your body belongs to the Führer). It is not unlikely that genetically enhanced persons would experience a similar double closure of their bodies. This is because, in a sense, their bodies would no longer belong to themselves but to the persons who designed them, meaning they would 'be' a body that they don't 'have,' but also because – and this constitutes the second closure – their bodies could also be said to belong to science and the free market, for it is at this level that decisions would be made about which genetic traits and characteristics were desirable or not. With regard to Esposito's third Nazi immunitary *dispositif*, the "anticipatory suppression of birth,"[127] we should note immediately that the interventions proposed by liberal eugenics are far less intrusive than those of Nazi eugenics. With genetic intervention, selecting the types of human beings to bring into existence would no longer occur through enforced sterilization or euthanasia, but by endowing embryos directly with the desired genetic traits. Note, however, that a ban on genetic intervention, as proposed for exam-

[123] Esposito, *Bíos*, 138-140.

[124] *Ibid.*, 141-143.

[125] *Ibid.*, 141.

[126] *Ibid.*, 142.

[127] *Ibid.*, 143-145.

ple by Habermas and Fukuyama, would also involve a kind of selection or, in negative terms, "suppression of birth," since this would imply that certain enhanced forms of life would not be allowed to come into being.

This brings us to Agamben's take on the biopolitics of genetic intervention. It should be noted that Agamben himself has never explicitly intervened in the liberal eugenics debate; however, there is some evidence in his writings, most notably in *Homo Sacer*, that he believes that the production of bare life is now most likely to occur in the field of medical intervention. For example, having analyzed Karl Binding's book *Authorization for the Annihilation of Life Unworthy of Being Lived*, discussed earlier in this chapter, Agamben suggests that this category may easily be extended beyond the limits set by Binding's discussion and, moreover, argues that:

> It is even possible that this limit, on which the politicization and the *exceptio* of natural life in the juridical order of the state depends, has done nothing but extend itself in the history of the West and has now – in the new biopolitical horizon of states with national sovereignty – moved inside every human life and every citizen. Bare life is no longer confined to a particular place or a definite category. It now dwells in the biological body of every living being.[128]

Agamben remains, however, extremely vague about the concrete implications of this supposed intensification and generalization of the production of bare life in the contemporary world. He only says that the line marking the point at which biopolitics can turn into thanatopolitics is "now in motion and gradually moving into areas other than that of political life, areas in which the sovereign is entering into an ever more intimate symbiosis not only with the jurist but also with the doctor, the scientist, the expert, and the priest."[129] Remember, however, that what Binding's book argues is that the law on suicide should be extended to the killing of the incurably ill and the mentally disabled – hence to "life unworthy of being lived." Agamben takes this as a statement that the "sovereignty of the living man over his own life has its immediate counterpart in the determination of a threshold beyond which life ceases to have any juridical value and can, therefore, be killed without the commission of a homicide."[130] We can well imagine, then, what Agamben might have to say about the project of a liberal eugenics. If parents' reproductive freedom were to be legally extended to include the right to intervene in the genetic makeup of their offspring, then this would cause a similar shift in

128 Agamben, *Homo Sacer*, 139-140.

129 *Ibid.*, 122.

130 *Ibid.*, 139.

the threshold of harm below which life would be seen as "wrongful" or "valueless," and thus not worth living.

Yet, Agamben's analysis of biopolitics also illuminates another crucial aspect of the liberal eugenics debate. Earlier, we mentioned that Scholem and Benjamin's discussion of the inscrutability of the law in Kafka's writings was important to Agamben's understanding of the perfect coincidence of law and life in the state of exception. In particular, it was Benjamin's view that, in Kafka's writings, the law seems to have become indistinguishable from life that proved to be a key inspiration for Agamben's work on the intimate connection between biopolitics and juridico-sovereign power. Perhaps we could say, then, that the debate over the moral and legal acceptability or otherwise of genetic technologies first makes it clear what he means by this. After all, whether or not these technologies will be permitted in the future, it will certainly be the case that the very existence or non-existence of certain potential forms of life will depend completely on a politico-juridical decision: in the coming age of genetic engineering, life might possibly become really indistinguishable from law.

How, then, should philosophy respond to the bleak prospect of a future in which the biological existence of human beings is increasingly contingent upon the twin power of juridico-political regulation and technological development? As we have argued, the dangers associated with a project such as liberal eugenics cannot be removed simply by placing legal restrictions on the further development of genetic technologies, nor by instituting "a right to a genetic inheritance immune from artificial intervention." Such conservative responses to the question of genetic intervention will not avert these dangers, but in fact reproduce them on another plane. Both Agamben and Esposito have therefore suggested that in order to escape the perils of contemporary biopolitics, it is urgently necessary to develop a conception of a life-affirming politics that can no longer be reversed into a life-negating politics. In Agamben's words:

> [U]ntil a completely new politics – that is, a politics no longer founded on the *exceptio* of bare life – is at hand, every theory and every praxis will remain imprisoned and immobile, and the "beautiful day" of life will be given citizenship only either through blood and death or in the perfect senselessness to which the society of the spectacle condemns it.[131]

Unfortunately, Agamben provides no comprehensive description of what this new politics might look like.[132] But he does make it clear that it

131 *Ibid.*, 11.

132 Although we will not go deeper into this matter here, it should be noted that, in the past couple of years, Agamben has increasingly argued that the foundations of such a

will require us to seek a politico-philosophical redefinition of life. Or, as Esposito puts it, the emergence of a more responsible biopolitics or what he calls "affirmative biopolitics" will depend on our efforts to "interpret life's relationship with politics philosophically."[133] In *Means Without Ends*, Agamben defines this new conception of life as "form-of-life:"

> By the term *form-of-life*, I mean a life that can never be separated from its form, a life in which it is never possible to isolate something such as naked life. [...] It defines a life – human life – in which the single ways, acts, and processes of living are never simple *facts* but always and above all *possibilities* of life, always and above all power. Each behavior and each form of human living is never prescribed by a specific biological vocation, nor is it assigned by whatever necessity; instead [...] it always retains the character of possibility; that is, it always puts at stake living itself.[134]

The distinctive characteristic of this new conception of life is that it no longer allows for a separation between biological life and political life, *zoē* and *bíos*. It is therefore a form of life in which it is never possible to isolate bare life as an object of sovereign power. In other words, form-of-life is a life in which the biological and political dimensions of the human being coincide without any remainder and is therefore "a life over which sovereignty and right no longer have any hold."[135]

In the remaining chapters we will seek to explore and substantiate this claim in more concrete terms. While remaining faithful to both Agamben's and Esposito's notions of affirmative biopolitics, our approach differs from theirs in two ways. First, we will show that Esposito is mistaken in suggesting that the outlines of an affirmative biopolitics premised on form-of-life cannot be found in the work of Hannah Arendt because, as he says, "Arendt didn't think the category of life thoroughly enough."[136] I will show in Chapter Four that it was not Arendt who "reached a blind spot concerning the problem of biopolitics,"[137] but that Esposito himself did not read Arendt carefully enough. It will be argued that, contrary to the standard interpretation of her work – including Esposito's – Arendt

---

"coming politics" will have to be sought in forms of religious anarchism, especially in the Paulinian messianic tradition. See for example Giorgio Agamben, *The Time that Remains: A Commentary on the Letter to the Romans*, trans. P. Dailey (Stanford: Stanford University Press, 2005).

133 Esposito, *Bíos*, 150.

134 Giorgio Agamben, *Means Without Ends: Notes on Politics*, trans. V. Binetti and C. Casarino (Minneapolis: University of Minnesota Press, 2000), 3-4.

135 *Ibid.*, 115.

136 Esposito, *Bíos*, 150.

137 *Ibid.*

didn't argue for a return to the ancient Greek understanding of politics with its strict separation of the private and the public, *oikos* and *polis*. On the contrary, her concept of *natality* precisely undoes any strict division between the realms of freedom and necessity because it names the radical co-implication of biological birth and politico-linguistic birth, *zoē* and *bíos*. Second, we will challenge Agamben's claim that "the biopolitical body of the West [...] cannot be overcome in a passage to a new body – a technical body."[138] In Chapter Five, we will argue, through a reading of the work of Bernard Stiegler, that it makes no sense to speculate about the biopolitical significance of a possible *passage* to a new technical human body, because the human and the technical have always been radically bound up with each other. To conclude, we will suggest that this notion of the "originary technicity" of life throws fresh light on the genetic enhancement debate. First, however, we will turn to Heidegger's dispersed writings on biotechnology and the question of genetic intervention. It will be suggested that his understanding of these developments is far more original and fruitful than those put forward by the participants in the current debate but that, despite his critical re-articulation of key concepts such as 'life,' 'nature,' and 'technology,' he ultimately fails to rethink the question of genetic enhancement in a sufficiently radical way.

[138] Agamben, *Homo Sacer*, 188.

Chapter Three

Enframed Life[1]

> Τέχνη can merely cooperate with φύσις, can more or less expedite the cure; but as τέχνη it can never replace φύσις and in its stead become the άρχη of health as such. This could only happen if life as such were to become a technically producible artifact. However, at that very moment there would also no longer be such a thing as health, any more than there would be birth and death. Sometimes it seems as if modern humanity is rushing headlong towards this goal of producing itself technologically. If humanity achieves this, it will have exploded itself, i.e., its essence qua subjectivity, into thin air, into a region where the absolute meaningless is valued as the one and only 'meaning' and where preserving this value appears as the human 'domination' of the globe.[2]

Heidegger and Biotechnology

Heidegger wrote down these prophetic words as early as 1939. One must, therefore, admire his lucid appraisal of the possibilities lying dormant in the scientific project of his time, the realization of which must have seemed but the most paradigmatic case of science-fiction. It is true that considerable progress had already been made in the domain of what we today call the life-sciences, which probably allowed Heidegger to engage in these speculations about the advent of a biotechnological revolution in the first place. For example, in his essay 'Overcoming Metaphysics' Heidegger refers to the pioneering research of the 1938 Noble Prize winner in chemistry Richard Kuhn, which, as Heidegger suggests, "already opens up the possibility of directing the breeding of male and female organisms according to plan and need,"[3] and he adds that "[s]ince man is the most important raw material [*Rohstoff*], one can reckon with the fact that someday factories will be built for the artificial breeding of human

[1] An earlier version of this chapter was published as Nathan Van Camp (2012), 'Heidegger and the question concerning biotechnology,' *Journal of Philosophy of Life* 2 (1), 32-54.

[2] Martin Heidegger, *Pathmarks*, ed. W. McNeill (Cambridge: Cambridge University Press, 1998), 197.

[3] Martin Heidegger, *Four Seminars*, trans. A. Mitchell and F. Raffoul (Bloomington: Indiana University Press, 2003), 106.

material [...]."[4] Furthermore, in his Heraclitus seminar with Eugen Fink, Heidegger discusses the coercive steering of genes in cybernetic biology.[5] And in one of his Le Thor seminars, he argues that what is most disturbing today is the "transformation of biology into biophysics," which will bring about that "the human can be produced according to a definite plan just like any other technical object."[6]

With the advantage of more than seventy years of hindsight, we know that genetic science still cannot actually industrially produce "human material," let alone entire human beings, but it would be unfair to simply brush aside Heidegger's speculations on the basis of his clearly oversimplified description of the actual technical procedures involved. Indeed, we have meanwhile witnessed the emergence of technologies that allow us to manage reproduction such as IVF and artificial insemination and while it is already possible to transform embryonic stem cells into other kinds of cell tissue for therapeutic purposes, it will very soon probably also be possible to grow stem cells into complete organs that can be used to replace damaged ones. While the curative potential of these technologies is almost unanimously hailed as a positive achievement of contemporary techno-science, this is certainly much less the case with technologies for reproductive cloning and genetic enhancement. The possibility to intervene in the human genome and to improve the constitution of the human body and mind inevitably raises questions not only about its technical feasibility, but also and especially about its moral permissibility. As said, the dangers the opponents of genetic enhancement invoke to prevent these technologies from being further pursued are manifold. They range from the increased risks of deformations and miscarriages that might result from experimentation on humans and, later on, from the technology's actual implementation, to the prospect of a world populated by a uniform type of human beings, to the fear that mankind will be divided into a small enhanced elite and a large non-enhanced underclass. They therefore urge governments to adopt regulations that bring the development of intrusive biotechnologies to a halt. The main question, however, is whether it is actually possible to distinguish between the appropriate and inappropriate use of biotechnology.

After recapitulating the debate between Jürgen Habermas and the advocates of a liberal eugenics about the moral permissibility of human genetic enhancement, it will be argued that both sides share a number of

4 *Ibid.*

5 Martin Heidegger and Eugen Fink, *Heraclitus Seminar*, trans. C. H. Seibert (Evanston: Northwestern University Press, 1993), 12-14.

6 *Ibid.*, 55.

assumptions about ideas of 'life,' 'nature' and 'technology' that prevent them from reflecting on the challenge that is presented by the current biotechnological revolution in a sufficiently radical way. Subsequently, we will turn to Heidegger's writings and suggest that his philosophy of technology allows us to think the question of biotechnology in a more fruitful way, but that, ultimately, he takes the radical challenge of biotechnology both too seriously and not seriously enough. Too seriously, because it is unclear why he would fear the annihilation of man's essence in a future biotechnological epoch if he is convinced that this essence is not related to man's biological equipment in the first place. Not seriously enough, in the sense that by arguing that the danger of biotechnology is that it challenges us to *understand* ourselves as modifiable organic material, Heidegger at the same time remains convinced that even the most intrusive interventions in the human body and mind will not disrupt man's ontological essence.

The Grown and the Made

One of the side-effects of the current biotechnological revolution is that the literary genre of science-fiction has suddenly gained an unexpected credibility. Aldous Huxley's *Brave New World* remains exemplary in this respect. Although many of such literary speculations about humanity's biotechnological future are dismissed by scientists as products of an imagination gone wild, there is every reason to take dystopian fictions like these seriously. After all, most of the technologies Huxley described in his novel, like the use of psychotropic drugs and artificial insemination, are currently available while others, like genetic engineering, will very likely become available in the near future. In *The Future of Human Nature*, Jürgen Habermas gives two additional reasons why we should pay close attention to even the most speculative predictions of this kind. First, there is always the risk that the ever increasing speed of scientific discoveries will take society by surprise and render it impossible for decision making bodies to put a stop to certain developments once they are on the verge of being implemented in clinical settings. Second, the mere expectation that more far-reaching possibilities will become available in the not so distant future has already brought about a radical uprooting of "deep-rooted categorical distinctions which we have as yet, in the description we give of ourselves, assumed to be invariant,"[7] and the awareness of this alone "might change our ethical self-understanding as a species in a way that could also affect our moral consciousness."[8]

7 Habermas, *The Future of Human Nature*, 42.

8 *Ibid.*

From a political perspective, decision making bodies are confronted with the paradoxical challenge that they have to enact clear cut regulations regarding technologies that at the same time radically destabilize the conceptual distinctions upon which our normative frameworks are grounded in the first place. Nowhere is this destabilization more disquieting, Habermas suggests, than in the case of the hitherto obvious distinction between "the nature we *are* and the organic equipment we *give* ourselves,"[9] between "what is *manufactured* and what has *come by nature*,"[10] or in its ancient Greek formulation, between entities which are engendered by *tekhnē* and those which are engendered by *physis*. As said earlier, the ethical significance of the distinction between the grown and the made is that it is intrinsically connected with the morally proper modes of action one should adopt when dealing with entities pertaining to one of these two ontological realms. In contrast to inorganic entities, whose inert nature makes them freely available for technical-instrumental interventions, the auto-regulated nature of organic entities severely restricts the modes of action one should impose on them. Referring to Aristotle's description of the various attitudes one should adopt when dealing with organic entities, Habermas gives the examples of the peasant, the doctor and the breeder who should respectively cultivate, heal, and select the entities which are entrusted to their care. Only these 'therapeutic' modes of action are responsive to the inherent dynamics of self-regulated organic entities. Since Habermas ultimately wants to argue that "the genetic foundations of our existence should not be disposed over,"[11] he thinks that only a revaluation of an Aristotelian distinction between different modes of action could yield specific guidelines to determine where the boundary lies between a negative eugenics, which Habermas considers permissible and even ethically binding, and a positive eugenics, which he thinks should be legally banned. The difference between a negative eugenics and a positive eugenics is that in case of the former genetic interventions are carried out to prevent or cure genetic defects, while in case of the latter they are aimed at enhancing man's genetic make-up. Therefore, in Habermas's view, the problem is not human genetic modification per se, but its use beyond the traditional logic of healing.

While the vivid memory of Nazi Germany's experimentation with eugenic policies probably explains why Habermas is so determined to set out clear boundaries between a positive eugenics and a negative eugenics, in the Anglo-Saxon world the discussion about genetic enhancement

9 *Ibid.*, 22.

10 *Ibid.*, 46.

11 *Ibid.*, 22.

has taken a different direction. Here, the main issue is to show that in contrast to the 'old eugenics' or 'authoritarian eugenics,' the 'new eugenics' or 'liberal eugenics' holds out the promise to drastically increase the individual's freedom to design how her future and that of her offspring will look like. As one of main proponents of the liberal eugenics, Nicholas Agar, argues, while the old eugenics "sought to produce citizens out of single centrally designed mould, the distinguishing mark of the new liberal eugenics is state neutrality."[12] In Agar's view, the drama of 20th century eugenics was that it was a state directed project which sought to coercively improve the genetic stock of the population in the direction of a 'healthier' or 'superior' race. In contrast, proponents of a liberal eugenics think that it should not be centrally decided which direction these improvements should take, but that it is up to the free individuals themselves to choose which genetic traits they want to provide their children. Agar asks, for example, whether Habermas's distinction between a negative eugenics and a positive eugenics does not also unwittingly reproduce certain politico-normative presuppositions. It seems indeed an impossible and even a dubious task to draw up an unbiased list of physical and psychical characteristics which should be considered defective and which, in Habermas's view, would therefore be eligible for genetic therapy. In Agar's view, the inability to draw a solid boundary between healing and enhancing is paralleled by the inability to make a clear distinction between a positive eugenics and enhancement of the human by modifications of the environment. For example, in Agar's view there is no difference between increasing a person's intelligence by genetic intervention or by providing her with an excellent education and that opponents of genetic enhancement often overestimate the role of genetic factors and downplay the role of environmental factors in the shaping of a person.

Agar's argument that there is no substantial difference between genetic enhancement and socialization obviously poses a problem for Habermas who maintains that there is a fundamental difference between the manipulation of 'subjective nature' and the manipulation of 'objective nature.' Yet, Habermas argues that the advocates of a liberal eugenics all too easily ignore the embodied perspective of the programmed person herself and do not take into account that genetically modified humans would be refused the possibility to respond to the demands that are imposed on them by their creators. Knowledge of the fact that one's dispositions, talents and bodily characteristics are genetically programmed will inevitably have consequences for one's self-understanding as it will transform intergenerational communication into an unidirectional imposition.

---

12 Agar, *Liberal Eugenics*, 137.

While the parents' efforts to modify the environment of their child to steer her development into a certain direction might be subsequently met by the child by developing a revisionary self-understanding, the possibility to change one's life history is cut off for the child when it concerns a modification of the genome.

Slavoj Žižek has, however, pointed out a striking paradox in Habermas's reasoning. If Habermas fears that modification of the human genome will decisively alter the fundamental parameters guiding the human's self-understanding and, more specifically, that it will deprive the human being of her inalienable right to possess a genetic inheritance immune from intervention by another person, then he is basically arguing that human beings can only retain their freedom and autonomy by leaving the distribution of genetic dispositions and characteristics to the contingent processes of nature. The paradox obviously being that, henceforth, "autonomy can only be maintained by prohibiting access to the blind natural contingency that determines us, that is, ultimately, by limiting our autonomy and freedom of scientific invention."[13] What would, in Habermas's view, deprive a person of her freedom, Žižek argues, "is, paradoxically, the very fact that was hitherto left to chance becomes dependent on the free decision of another person."[14] Habermas's argument that human beings can only experience their freedom with reference to something which is not at their disposal only makes sense if this gifted aspect is *really* out of their reach. If this is not the case, however, then it seems that Habermas is actually arguing that the only way for us to maintain our sense of dignity and autonomy is by committing ourselves to an illusion.

When asked in the posthumously published *Der Spiegel* interview what is exactly wrong with technology given the many benefits it offers to mankind, Heidegger indignantly replied:

> Everything is functioning. This is exactly what is so uncanny, that everything is functioning and that the functioning drives us more and more to even further functioning.[15]

In Heidegger's view, the danger of technology, including biotechnology, is not that something might go terribly wrong, but, quite the contrary, that everything will actually function according to plan. Heidegger would not deny that genetic intervention could provoke certain factual dangers,

---

13 Slavoj Žižek, *Organs Without Bodies* (London/New York: Routledge, 2004), 126.

14 Slavoj Žižek, *The Parallax View* (Cambridge Mass.: MIT Press, 2006), 307.

15 Martin Heidegger, *The Heidegger Controversy: A Critical Reader*, ed. Richard Wolin (Cambridge Mass.: MIT Press, 1993), 105-106.

but for him this is not what is most disquieting about it. The most threatening prospect would precisely be that the human genome could be modified without any undesirable side effects. In that case, Heidegger predicts, the human being would have transformed himself into an entity that could hardly still be called human. However, this does not mean that Heidegger would side with critics like Habermas and Bill Mckibben,[16] who argue in favor of a legal ban on germ line intervention, but do allow for somatic gene therapy. Since Heidegger rejects the broadly held view that technology is a means to an end, he considers the belief that one can control the development of technology through regulation a metaphysical illusion. Human beings do not master technology. Just when they believe they can control and regulate technology, it may turn out that they are actually its most docile servants.

Now, if technology is not merely a means to an end, then, by challenging the de-differentiation of eugenics and socialization on the argument that it erroneously "presupposes a leveling out of the difference between the grown and the made, the subjective and the objective,"[17] Habermas seems to succumb to a curious perspective fallacy, for it seems that this kind of intervention will precisely render meaningless the distinction between what is 'natural' and what is 'artificial.' In a biotechnological future, this distinction will be performatively abolished, so to speak, and it would therefore be self-deceptive to continue to use this distinction as a point of reference for evaluating the totally new reality that is looming on the horizon. As Slavoj Žižek puts it:

> [D]oes the very fact of the possibility of biogenetic manipulations not retroactively change the self-understanding of ourselves as "natural" beings, in the sense that we now experience our "natural" dispositions themselves as something "mediated," not simply as something immediately given but as something that can be in principle manipulated (and is thus simply contingent)?[18]

To argue that genetic intervention will destroy the 'naturalness' of the human presupposes the idea that human beings had a 'natural' essence in the first place. Yet, this assumption is precisely what the possibility of genetic modification radically puts into question and, therefore, cannot be wielded as an argument against it.

The advocates of a liberal eugenics, on the other hand, start from the same *petitio principii*, but arrive at the exact opposite conclusion. Since

---

[16] Bill Mckibben, *Enough! Staying Human in an Engineered Age* (New York: Henry Holt & Company L. L. C., 2003), 127.

[17] Habermas, *The Future of Human Nature*, 50.

[18] Žižek, *Organs Without Bodies*, 125-126.

they also conceive technology as a means, they expect to find the human being unscathed on the other side of the biotechnological threshold. They assume that genetically enhanced humans will have the same goals, aspirations, and dreams as their non-enhanced predecessors, but with this decisive difference, that they will be better equipped to achieve them. Heidegger already exposed this fallacy as early as 1967 in a lecture for the Greek Academy of Arts and Sciences.[19] The essence of modern science, Heidegger argues here, reveals itself in the victory of the scientific method over science, because it is the method that determines in advance which phenomena are experimentally accessible and which ones not. On Heidegger's account, the victory of the scientific method manifests itself pre-eminently in what was at the time considered to be the science of the future, namely cybernetics. Cybernetics – from the Greek *kybernētēs*, steersman – only takes into account the reality of information flows between various systems and subsystems. Significantly, it does not differentiate between organic and inorganic systems. In cybernetics, it is assumed that both are steered by the same basic principle, namely the feedback loop of information. It is by applying these principles to the study of the human body, Heidegger argues, that the life sciences discovered that the germ cell contains the information that regulates the living character of human beings [*Das Lebendige im Leben des Menschen*], their "life plan" [*Lebensplan*], so to speak. It is by gaining access to this kind of information that "one day we will be able to techno-scientifically produce and breed human beings."[20] However, the problem with the biotechnological future as projected by what Heidegger calls "futurology" [*Futurologie*] is that it is merely a "stretched present" [*eine verlängerte Gegenwart*]. Biotechnology remains a project of the epoch of subjectivity, but it curiously enough ignores the fact that its accomplishments will actually nullify the distinction between subject and object.

Dasein and Life

Perhaps the most fundamental weakness of all arguments against genetic enhancement is that they accept as given the idea that the human being's essence is ultimately grounded its genetic constitution. From this point of view, it can indeed be reasonably argued that genetic intervention will dramatically change our *present* understanding of what a meaningful

19 Martin Heidegger, *Denkerfahrungen 1910-1976* (Frankfurt a/M: Klostermann Verlag, 1983), 135-149.

20 "Aus seine Kenntnis gründet man die sichere Aussicht, eines Tages die wissenschaftlich-technische Herstellbarkeit und Züchtung des Menschen in der Griff zu bekommen" (*ibid.*, 143).

existence entails, yet, by the same token, it cannot be maintained that it will deprive existence of meaning *as such*. As Žižek remarks: "Who knows what this 'posthuman' universe will reveal itself to be 'in itself'?"[21] Heidegger is an exception in this regard. Although he also regards human genetic modification as an imminent danger for the human essence, he nonetheless consistently rejected biological determinations of this essence. He was never tired of repeating that the metaphysical tradition erroneously defined the human being as a natural entity to which some higher capacity such as speech, reason, or soul is added. According to Heidegger, the essence of the human being lies in his receptivity to Being, which he calls "being-in-the world," "ek-sistence," or "transcendence:"

> Ek-sistence can be said only of the essence of man, that is, only of the human way 'to be.' For as far as our experience shows, only man is admitted to the destiny of ek-sistence. Therefore ek-sistence can also never be thought of as a specific kind of living creature among others.[22]

Does that mean that according to Heidegger the human being does not belong to the realm of nature, or even that the human being is not a living creature at all? Such a position would even exceed religious dogma in anti-Darwinism. Yet, for Heidegger the burden of proof does not rest with those who, like himself, challenge what he calls "zoological definitions" of the human, but with those who claim that the essence of the human being is somehow grounded in 'life.' The human being or Dasein is not 'life' plus something else added to it. Rather, 'life' is Dasein from which something is subtracted. As Heidegger puts it in *Being and Time*:

> Life, in its own right, is a kind of Being; but essentially it is accessible only in Dasein. The ontology of life is accomplished by way of a private interpretation; it determines what must be the case if there can be anything like mere-aliveness [*Nur-noch-leben*]. Life is not a mere Being-present-at-hand, nor is it Dasein. In turn, Dasein is never to be defined ontologically by regarding it as life (in an ontologically indefinite manner) plus something else.[23]

In *Being and Time*, Heidegger argues that Being-in-the world is Dasein's essential way of being, its most basic state. Hence, what needs to be subtracted from Dasein to arrive at an understanding of what 'life' means is 'world.' In *Being and Time*, Heidegger does not further investigate what this would mean. On the contrary, he argues that Dasein does

21 Žižek, *The Parallax View*, 195.

22 Martin Heidegger, *Basic Writings*, ed. D. F. Krell (New York: Harper & Row, 1977), 228.

23 Martin Heidegger, *Being and Time*, trans. J. Macquarrie and E. Robinson (New York: Harper & Row, 1962), 75.

not have access to such a thing as 'pure life' because our understanding of the entities that surround us is necessarily mediated by the world. The world constitutes the ultimate horizon of our understanding of entities behind which we cannot penetrate. The same can be said about the concept of 'pure nature.' The conception of natural beings which are somehow independently present as present-at-hand objects derives from a more originary understanding of entities that are ready-to-hand, namely tools and objects of use:

> The wood is a timber forest, the mountain a rock quarry, the river a hydraulic force, the wind 'in the sails.' When the 'environment' is disclosed, the 'nature' thus disclosed is encountered too.[24]

The same for animals. Although it is affirmed that animals are not simply artifacts or automatons, as Descartes still maintained, in Heidegger's existential analytic animals are entities that appear within the everyday world of Dasein as merely useful for this or that purpose: as leather for shoes, as power source, etc. The idea of a living being as such can only arise if one makes abstraction of this more original, worldly encounter.

Apart from a telling remark that animals simply perish, never properly die as Dasein does – of which Jacques Derrida will make much[25] – *Being and Time* remains otherwise silent on the question of the being of life. This should not come as too great a surprise, though. *Being and Time* was written with the explicit intention to reopen the question of Being as such. Therefore, the attempt to determine the being of particular regions of entities such as life can only be successful after this more fundamental issue has been properly addressed. Yet, merely three years later, in *The Fundamental Concepts of Metaphysics*, the question of animal being comes to occupy a central place in his thinking. The guiding threat of Heidegger's argument in this lecture course is constituted by three theses: the stone is wordless, the animal is world-poor and man is world-forming. Note that *Being and Time*'s conceptual framework still informs the relationship between these theses. Life is considered to be only accessible through a subtraction of world from Dasein: the animal is "poor in world." Hence, as Heidegger reasserts his earlier claim, "*[t]he essence of life can become accessible only if we consider it in a deconstructive fashion*."[26] Heidegger will conclude that the essence of the ani-

---

24 *Ibid.*, 100.

25 See Jacques Derrida, *Aporias*, trans. T. Dutoit (Stanford: Stanford University Press, 1993).

26 Martin Heidegger, *The Fundamental Concepts of Metaphysics: World, Finitude, Solitude*, trans. W. McNeill and N. Walker (Bloomington: Indiana University Press, 1995), 255.

mal consists in being "captivated" [*benommen*]. The animal is trapped in its environment by the ring or circle of its drives and instincts. The animal has access to the entities that penetrate its circle of instincts, but that it can only react to these entities, never truly act in relation to them. Heidegger avidly describes an experiment of the biologist Jacob von Uexküll, during which a bee is brought in front of a cup of honey. It appeared that the bee continued to suck up the honey even after the bee's abdomen was cut away and the honey visibly streamed away behind it. According to Heidegger, this experiment convincingly demonstrates that the bee does not recognize the food *as* too much food and that it can only continue its instinctual activity. What is withheld from the bee is the ability to apprehend the honey *as such*. Since the animal is completely captivated or absorbed by the entities constituting its environment, it cannot let an entity be what it is. This phenomenological capability to reveal an entity in its being is reserved for Dasein only. The fact that Dasein is world-forming implies that Dasein always already moves within a horizon of meaning (*Lichtung*) from out of which which present themselves *as* this or *as* that.

However, it remains unclear how Dasein's possibility to reveal entities as such relates to what is biologically given to him. Does it not find its condition of possibility there, in the sense that it surely cannot be denied that Dasein's ek-sistence or being-in-the-world must somehow be grounded in his physical constitution? This is not at all obvious to Heidegger. Quite the contrary, in fact. He maintains that ek-sistence is more original than the body and owes nothing to it. In sense perception, for example, we do not first register 'raw data' such as noises or sounds to which we subsequently apply the categories of understanding as to finally arrive at a meaningful representation. Such a hylomorphic detour is totally redundant. Phenomena immediately present themselves *as* this or *as* that:

> What we 'first' hear is never noises or complexes of sounds, but the creaking wagon, the motor cycle. [...] The fact that motor-cycles and wagons are what we proximally hear is the phenomenal evidence that in every case Dasein, as Being-in-the-world, already dwells *alongside* what is ready-to-hand within-the-world. [...] Dasein, as essentially understanding, is proximally alongside what is understood.[27]

On Heidegger's view, it is understanding which makes possible hearing, not the mere 'natural' fact of being endowed with ears. Heidegger maintains the priority of understanding over bodily sense perception to a point where it almost becomes absurd, as to even claim that faculties

[27] Heidegger, *Being and Time*, 207.

are more original than physical organs: "We do not hear because we have ears, but we have and are able to have ears because we can hear."[28] Biological organs merely register what, on a more fundamental level, has already been actualized by the faculty of understanding.

It should be more than clear, then, that Heidegger seeks at all costs to counter the modern tendency to define the essence of man in biological or physiological terms. Yet, even though such an approach could be plausibly defended, it seems to deprive his critique of biotechnology of any meaningful content, for why would he consider biotechnology as a threat to the human being's essence if this essence is thinkable in terms of the human's biological equipment in the first place? Such a critique seems to be in keeping with the reaction of the Catholic Church to cloning and genetic engineering: If the human being's essence, the soul, is really immortal, then why oppose biotechnological interventions on the perishable body? The answer is that Heidegger understands the notions of 'nature,' 'life' and 'technology' in a completely different way. Therefore, to really understand why Heidegger rejected the idea of human genetic modification, we must turn to his reinterpretation of the ancient Greek concepts of *physis* and *tekhnē*.

The Essence of Biotechnology

"Technology is not equivalent to the essence of technology,"[29] Heidegger states at the beginning of his famous essay 'The Question Concerning Technology.' Almost everything we thought we knew about technology – it is an inert, neutral means utilized by human beings – may well be "correct," but it is certainly not "true." To understand why these characteristics do not determine technology in its essence, Heidegger's claim has to be placed against the backdrop of his interpretation of the ancient Greek notion of *tekhnē*. *Tekhnē* does not refer to the production and utilization of tools and machines, but to a kind of knowledge, a know-how.[30] Heidegger explains this through an interpretation of Aristotle's theory of causality which, importantly, also seems to inform the still widely accepted conception of technology as a means. As is well known, Aristotle posits four causes: the material cause, the formal cause, the final cause, and the efficient cause. Conventional readings of Aristotle take this

28 Martin Heidegger, *Heraklit* (Frankfurt a/M: Klostermann Verlag, 1979), 247.

29 Martin Heidegger, *The Question Concerning Technology*, trans. W. Lovitt (New York: Harper & Row, 1979), 4.

30 "Technē means neither art not technology but a kind of knowledge, the knowing disposal over the free planning and arranging and controlling of arrangements." Martin Heidegger, *Introduction to Metaphysics*, trans. G. Fried and R. Polt (New Haven: Yale University Press, 2000), 18.

to imply that technology involves man as efficient cause imposing form upon inert matter to his own end. According to Heidegger, though, such an interpretation of the term is completely un-Greek. The term which we now call "cause," the Greeks called "aition," "that to which something else is indebted." This "something else" is the artifact, the end result of *tekhnē*. The moderns think that only the human being is responsible for the production of the artifact, in the sense that human beings are supposed to be those who control the production process from A to Z. The Greeks, however, understand this process differently. For the Greeks, the human being isn't the pivot in *tekhnē*, but merely one of the four *aitia*. The implication is that *tekhnē* is not a means to achieve a freely chosen end, but a process of co-creation involving all four *aitia* and leading to a coming into being of what is not yet a being. In Heidegger's words, *tekhnē* "lets what is not yet present arrive into presencing."[31] *Tekhnē* is a mode of *poiēsis*, a bringing-forth. This leads Heidegger to a final conclusion. As a mode of *poiēsis*, *tekhnē* lets something which is concealed come into unconcealment: *tekhnē* is closely related to what the Greeks called *alētheia*, truth. It is through the know-how involved in *tekhnē* that the human is able to grasp truth, which for Heidegger is nothing less than the truth of Being.

What about modern technology? Modern technology is also a mode of revealing, but no longer a bringing-forth, a *poiēsis* leading to truth, but a challenging-forth [*Herausfordern*] that "puts to nature the unreasonable demand that it supply energy that can be extracted and stored as such."[32] In modern technology, Being is reduced to a standing-reserve [*Bestand*] of energy that can be stockpiled and used to serve the human's subjective needs. Through modern technology humanity seems to be finally able to realize Bacon and Descartes' project of making itself master and possessor of nature. Heidegger casts this anthropocentric view into doubt. What appears to install humanity in a position of absolute mastery over the natural world is actually something that has been donated to it: we are only able to reveal Being as mere raw material ready for exploitation by modern technology because Being has already revealed *itself* to us in this form. In this sense, modern technology is not a human accomplishment, but a sending of Being. In Heidegger terms, it is a destining [*Geschick*]. Such a revelation of Being that challenges man to, in turn, challenge nature is what Heidegger calls *das Gestell* [*enframing*]. Now, not only does the human lack control over the revelation of Being at work within modern technology, but human beings are themselves profoundly subject to

31 Heidegger, *The Question Concerning Technology*, 10.

32 *Ibid.*, 14.

that revelation. Here we finally arrive at an answer to the question why Heidegger considers biotechnology a threat to the human essence:

> When destining reigns in the mode of Gestell, it is the supreme danger. [...] As soon as what is unconcealed no longer concerns man even as object, but does so, rather, exclusively as standing-reserve, and man in the midst of objectlessness is nothing but the orderer of the standing-reserve, then he comes to the very brink of a precipitous fall; that is, he comes to the point where he himself will have to be taken as standing-reserve.[33]

In the age of modern technology, and even more so in the dawning age of biotechnology, the human herself is challenged or revealed as a standing-reserve of genetic material to be exploited by the life-sciences. It should be noted, though, that the entire discussion is entirely played out on the level of understanding. The danger of biotechnology does not primarily consist in concrete empirical threats posed by technical interventions on the genome, but in the threat that we will *understand* ourselves as bio-genetic entities that can be manipulated at will. This is what is most dangerous about biotechnology, not the fact that we will actually use these technologies to modify the genome. This is why Heidegger would probably say that the discussion between Habermas and the advocates of a liberal eugenics about the permissibility of genetic intervention is beside the point. Both sides already presume that the human essence is somehow grounded in the human being's genetic constitution and, in that sense, the damage has already been done:

> The actual threat to man does not come in the first instance from the potentially lethal machines and apparatus of technology. The actual threat has already affected man in his essence.[34]

In Heidegger's view, such a tacit agreement between the advocates and the critics of genetic enhancement would merely demonstrate that the *Gestell* has become hegemonic and dictates the terms in which the discussion is to be pursued.

On the one hand, then, man is so thoroughly in the grip of the illusion of absolute sovereignty over Being conjured up by the *Gestell* that he is hardly able to recognize that it is still a mode of revealing of which he is merely the recipient:

---

33 *Ibid.*, 26-27.

34 *Ibid.*, 28.

> he does not apprehend *Gestell* as a claim. [...] he fails to see himself as the one spoken to, and hence also fails in every way to hear in what respect he ek-sists, from out of his essence, in the realm of an exhortation or address.[35]

On the other hand, the *Gestell*'s total monopoly over the process of revelation blocks off all other possible revelations of Being, including *poiēsis*:

> Where this ordering holds sway, it drives out every other possibility of revealing. Above all, the *Gestell* conceals that revealing which, in the sense of poiēsis, lets what presence come forth into appearance.[36]

In the end, then, the *Gestell* not only excludes all other possible revelations such as ancient *tekhnē*, but it even occults its own status as a revelation and the ontological structure of Being as a process of revealing. In other words, the *Gestell* simply becomes synonymous with "what is."[37]

Now, despite Heidegger's extreme pessimism regarding man's possibility to extract himself from the hegemony of the *Gestell*, he still mentions a "saving power." If the *Gestell* denotes Being in its contemporary form, then this means that it is also a revelation bestowed upon man, rather than one created or imposed by him. In fact, Heidegger already said at the beginning of the essay that it is only by thinking the "ambiguous essence"[38] of technology as a disclosure granted to the human by Being, rather than an empirical technical instrument in the hands of a free subject, that the human is able to receive the possibility of a "free relationship"[39] to modern technology. Since we know that for Heidegger the essence of technology is that it is a mode of revealing, the implication is that that such a free relationship can only be attained by retrieving the original meaning of *poiēsis*:

> Revealing is that destining which, ever suddenly and inexplicably to all thinking, apportions itself into the revealing that brings forth and that also challenges, and which allots itself to man. The Gestell has its origin as a destining in poiēsis.[40]

Why, however, after first arguing that the *Gestell* blocks *poiēsis*, does Heidegger now see the *Gestell* as itself poetic in origin and why, moreover, does he even credit it with containing the "saving power?"

35 *Ibid.*, 27.

36 *Ibid.*

37 *Ibid.*

38 *Ibid.*, 33.

39 *Ibid.*, 3.

40 *Ibid.*, 29-30.

At this point, it becomes necessary to turn to Heidegger's interpretation of the pre-Socratic notion of *physis*. *Physis* does not correspond to the domain of entities that we conventionally call 'nature,' which is actually merely a metaphysical interpretation of it, but names an ontological concept that is prior to any particular process that may take place like the growing of a tree, because it refers to the event through which particular entities originally receive their essence. In other words, *Physis* is nothing less than the original disclosure of Being itself. In the *Introduction to Metaphysics*, Heidegger says that "*physis* is Being itself, by virtue of which beings first become and remain observable."[41] More specifically, Heidegger defines *physis* as "what emerges from itself, the unfolding that opens itself up, the coming-into-appearance in such unfolding, and holding itself and persisting in appearance – in short, the emerging-abiding sway."[42] Furthermore, Heidegger explains that for the Greeks *physis* also included some of the concepts to which that notion is traditionally opposed like soul (*psychē*), position (*thesis*), law (*nomos*), and more importantly for our purposed, even technology (*tekhnē*). As Heidegger will repeat in 'The Question Concerning Technology:'

> [N]ot only handcraft manufacture, not only artistic and poetical bringing into appearance and concrete imagery, is a bringing-forth, *poiēsis*. *Physis* also, the arising of something from out of itself, is a bringing-forth, *poiēsis*. *Physis* is indeed *poiēsis* in the highest sense.[43]

If *poiēsis* is the *non*-technological essence of technology which we should retrieve from oblivion if we want to free ourselves from pervasive modern technology, it is because it achieves its highest form of expression in *physis*. Both *tekhnē* and *physis* are modes of *poiēsis*, but since *physis* is *poiēsis* to a higher degree, the poietic revealing of *tekhnē* can only be properly understood from the vantage point of *physis*. Heidegger explains why *physis* has priority over *tekhnē* in his essay on Aristotle's *Physics*. Faithful to Aristotle's basic principle that to know is to become one with the thing known, Heidegger seeks to explain the difference between *physis* and *tekhnē* by examining the things that are revealed through both modes of *poiēsis*: natural things (*physei onta*) and technical things (*tekhnē onta*). The most basic quality of a natural thing is that it is in movement (*kinesis*) or, in other words, that it has the tendency to change from something into something else (*metabolē*). Hence, for Aristotle, movement does not only mean change of place, but also growth

---

41 Heidegger, *Introduction to Metaphysics*, 15.

42 *Ibid*.

43 Heidegger, *Question Concerning Technology*, 10.

and diminution, alteration, and generation. Technical things are *also* moving beings, but usually we encounter them in a state of rest as that which has been produced and is now ready to use. Aristotle will thereupon argue that technical entities nonetheless differ from natural entities insofar as they move differently: while natural things contain within themselves the source (*archē*) of their movement, technical things have no impulse to change arising from themselves.

For the purpose of our present discussion we need not to go into the details of Heidegger's Aristotle interpretation. Let it suffice to say that the reason why Aristotle is such an important interlocutor for Heidegger is that his theory of *physis* is at the crossroads of the truthful account of the essence of *physis* in pre-Socratic philosophy and its later metaphysical determination that will culminate in the *Gestell*. In Heidegger's view, the moderns have turned the ancient Greek understanding of the relation between *physis* and t*ekhnē* upside down in the sense that in the age of the *Gestell* it is believed that it is in technology that *poiēsis* achieves its highest form of expression. In modernity, Being is no longer the "emerging-abiding sway" of *physis*, but is reduced to inert matter or standing-reserve passively awaiting technical formation by the human. Heidegger says, for example, that defining living beings as organisms actually boils down to interpreting them as "artifacts that make themselves."[44]

Aristotle could be said to have inspired such a technological understanding of Being because he tried to explain the self-emerging sway of *physis* in terms of matter and form. Heidegger therefore seeks to read Aristotle's form/matter distinction in such a way that it remains faithful to the pre-Socratic view that *physis* is superior to t*ekhnē*. Why is this so? First, because matter (*hylē*) and form (*morphē*) belong together necessarily in *physis*, but only incidentally in t*ekhnē*. Wood is necessary for a tree, but incidental to a table, which could also be made out of iron for example. Moreover, in contrast to the bringing-forth of technical things, in which the form must first exist in the mind of an external agent prior to production, in the generation of natural things no such external help is required. If they did require outside assistance, Heidegger remarks, "this would mean an animal could not reproduce itself without mastering the science of its own zoology."[45]

Second, since matter and form belong together only incidentally in *tekhnē*, technical things have no impulse to change arising from themselves, *except* insofar they are also always partly made out of natural things. A wooden bed will eventually rot not because it is a bed, but

44 Heidegger, *Pathmarks*, 195.

45 *Ibid.*, 222.

because it is made out of wood. The imposition of a form by an external agent in t*ekhnē* does not destroy the original complicity of form and matter in the natural things that form the basis of technical things. This is an important thing to note for Heidegger because it carries the implication that *physis* is necessarily prior to *tekhnē* and, therefore, that the movement characteristic of *physis* will always eventually overpower the movement of t*ekhnē*. For the ancients, then, *tekhnē* does not consist in imposing a preconceived form on inert matter, but in a mode of poietic disclosure that operates in accordance with the poietic *self*-disclosure of *physis*:

> *Tekhnē* is a mode of proceeding against *physis*, though not yet so as to empower it or exploit it, and above all not to turn use and calculation into principles, but, on the contrary, to retain the holding sway of *physis* in unconcealment.[46]

The human is not the master of nature, but merely assists the embryonic process of coming-into-appearance that is already underway in *physis*. Such an understanding of technology and nature sounds quite alien to modern ears. It implies, for example, that the production of a wooden table doesn't involve a carpenter imposing a subjectively chosen shape to an indifferent piece of wood, but that the wood, the *hylē*, already contains the potential to disclose itself as a table and that the carpenter merely 'assists' the wood to show itself as what it potentially already is:

> The change of the appropriate wood into a table consists in the fact that the very appropriateness of what is appropriated emerges more fully into view and reaches its fulfillment in the appearance of a table and thus comes to stand in the table that has been pro-duced, placed *forth*, i.e., into the unhidden.[47]

Now we are sufficiently prepared to return to the fragment we quoted at the beginning of the chapter:

> *Τέχνη can merely cooperate with φύσις, can more or less expedite the cure; but as τέχνη it can never replace φύσις and in its stead become the ἀρχη of health as such.*

It is not the doctor or the engineer who cause health by their technical skills, but the emerging-abiding sway of *physis*. Medicine or biotechnology merely determine the conditions for the body's self-emergent health to reassert itself, but ultimately it is always *physis* which heals.

46 Martin Heidegger, *Basic Questions of Philosophy: Selected Problems of Logic*, trans. R. Rojcewicz and A. Schuwer (Bloomington: Indiana University Press, 1994), 179-180.

47 Heidegger, *Pathmarks*, 218.

Sometimes it seems as if modern humanity is rushing headlong towards this goal of producing itself technologically. If humanity achieves this, it will have exploded itself, i.e., its essence qua subjectivity, into thin air [...].

If the human decides not to follow the leads that *physis* provides and declares herself to be the master of *physis*, then humanity is on the verge of losing its essence. Note, though, that Heidegger says that the human will then have exploded his essence qua *subjectivity*. Humans will no longer understand themselves as autonomous subjects because they will become aware that they are much less placed in a situation of mastery over nature by technology than they are subjected to the imperatives of technology. It is for this reason that Heidegger claims that the saving power only arises when the danger is at its highest.

But can human beings also lose their ontological essence qua *ek-sistence*? Can they abandon their relation to Being? It seems not, for, as we have seen, also as the servant of technology they merely respond to the call of Being that claims them to challenge Being as standing reserve. Even if they would be completely genetically modified, this would not really affect the core of their being because their essence is not grounded in their biological constitution, but in their ek-static belonging to Being. What is uncanny about technology, Heidegger says, is that "calculative thinking may someday come to be accepted and practiced as the only way of thinking."[48] Yet, modern technology as calculative thinking is still a mode of thinking and as such remains responsive to the self-revealing of Being.

In the final analysis, then, it appears that, in Heidegger's view, bio-technology is not so dangerous after all. It might well be the case that the life-sciences may one day succeed in reducing man to a homeostatic biogenetic entity, but for Heidegger one thing remains certain: "Man will never become a machine."[49] The human's ek-static essence renders such a transformation simply impossible. The use of biotechnology could temporarily suppress the emerging-abiding sway of *physis*, but in the long run *physis* will always prevail. In a sense, then, Heidegger is very close to Habermas who also argues that what is at stake in biotechnology is the blurring of the boundaries between the grown and made. But in contrast to Habermas, Heidegger remains convinced that the human will never lose the ability to discriminate between what is engendered according to *physis* and what is engendered according to *tekhnē*:

---

48 Martin Heidegger, *Discourse on Thinking*, trans. J. M. Anderson and E. H. Freund (New York: Harper Perrenial, 1966), 56.

49 Martin Heidegger, *Bremer und Freiburger Vorträge* (Frankfurt a/M: Klostermann Verlag, 1994), 37.

> Differentiating between what appears of and by itself from what does not appear of and by itself is a χρίνιν [*krinein*, to decide or to separate] in the genuinely Greek sense: separating out what is superior from what is inferior. Through this "critical" ability for differentiating, which is always decision, the human being is lifted out of mere captivation by what presses upon and preoccupies him or her and is placed out beyond it, into the relation to being.[50]

In other words, the human can neither lose the potential to discriminate between the grown and the made, nor the potential to recognize that the former is far superior to the latter, because the human actually *is* nothing other than this potential, that is to say, the entity which exists in the form of *potentiality-for-being*.

[50] Heidegger, *Pathmarks*, 202.

CHAPTER FOUR

Natal Life[1]

> History and nature have become equally alien to us, namely, in the sense that the essence of man can no longer be comprehended in terms of either category.[2]

Hannah Arendt and Biotechnology

In what are practically the opening lines of *The Human Condition* Hannah Arendt makes a bold yet apt prediction about the future of humanity:

> For some time now, a great many scientific endeavors have been directed toward making life also "artificial," toward cutting the last tie through which even man belongs among the children of nature. It is the same desire to escape from imprisonment to the earth that is manifest in the attempt to create life in the test tube, in the desire [...] to produce superior human beings and to alter their size, shape and function.[3]

Given the wide interest in contentious bioethical issues like reproductive cloning and genetic enhancement, it is somewhat surprising that so few have yet taken serious interest in this intriguing remark. This is as much striking as is it is regrettable, not only because Arendt's thought could throw a fresh light on a prominent question in much current bioethical and politico-philosophical debates, but also because the specific stakes of the question of biotechnology could, in turn, yield a new understanding of her celebrated yet still rather poorly understood concept of natality. "Natality," Arendt explains at the beginning of *The Human Condition*, "may be the central category of political thought," because "the new beginning inherent in birth can make itself felt in the world only because the newcomer possess the capacity of beginning

[1] An earlier version of this chapter has been published as Nathan Van Camp (2014), 'Enhancing the natal condition: Hannah Arendt and question of biotechnology,' *Symposium: Canadian Journal of Continental Philosophy* 18 (2), 179-189.

[2] Hannah Arendt, *The Origins of Totalitarianism* (New York: Schocken Books, 2004), 378.

[3] Arendt, *The Human Condition*, 2.

something anew, that is, of acting."[4] Thus, if the human capacity to act is ontologically rooted in the bare biological fact of being born, then it seems that prenatal technological intervention in the genetic endowment of a human being cannot remain without consequences for the newborn's capability to become a free acting agent. It is hardly surprising, then, that the few authors who do invoke Arendt's thought in the context of an ethical discussion of biotechnology do so with specific reference to the concept of natality.[5]

Yet, despite Arendt's clear concern about the possible effects of such technological developments on the human capacity to initiate new beginnings, it is far from obvious that Arendt's concept of natality lends itself so obviously to arguments that oppose genetic intervention, as, for example, Habermas and Michael Sandel seem to believe. It is indeed true that she sees in the desire to redesign our biogenetic base a "wish to escape the human condition" and a "rebellion against human existence as it has been given,"[6] but that does not necessarily mean that she also considered it a threat to the human's natal condition. Rather, it may even be surmised that the opposite is true, that is, that she developed the concept of natality precisely in response to this looming possibility. Habermas invokes the condition of natality to argue that, for human freedom to be retained, it is essential that our biogenetic properties must remain outside the reach of planned technological intervention. But is it not paradoxical to argue that human freedom is grounded in something that is given by necessity? Readers of Arendt have always been puzzled by the fact that, while she appears to assume a sharp distinction between the realms of necessity and freedom, she nonetheless argues that the capacity to act freely is rooted in the biological fact of birth. In this chapter it will be argued that this apparent contradiction can only be removed if we recognize that natality is a concept that deconstructs any blunt opposition between *zoē* and *bios*, biological birth and politico-linguistic birth. Expanding on Arendt's largely neglected footnote to the German anthropologist Arnold Gehlen in *The Human Condition*, it will be argued that the key to this enigma can be found in what is known in evolutionary biology as the phenomenon of neoteny or fetalization. It will become clear, then, that Habermas and

---

4 *Ibid.*, 9.

5 See Habermas, *The Future of Human Nature*, 58-60; Sandel, *The Case Against Perfection*, 82-83; Anne O'Byrne, Natality *and* Finitude (Bloomington: Indiana University Press, 2010) and Jean-Pierre Dupuy, 'Cybernetics is Antihumanism: Advanced Technologies and the Rebellion Against the Human Condition,' in *Transhumanism and its Critics*, ed. G. R. Hansell and W. Grassie (Philadelphia: Metanexus, 2011), 227-248.

6 Arendt, *The Human Condition*, 2.

Sandel are mistaken to assume that her concept of natality yields a strong argument against human genetic intervention.

The Techno-Human Condition

An entry in Arendt's *Denktagebuch* (Thought Diary) dated early 1958 suggests that one of the main reasons why she abandoned the project of writing a book on the totalitarian elements in Marxism and instead began to focus on the basic components of the *vita activa*, was that she became increasingly occupied with the question of what impact major impending technological developments will have on the human condition:

> Ad Kafka, "Investigations of a Dog:" The greatest and most horrible danger for human thought is that what we once believed to be true could be shattered by the discovery of a fact that was previously unknown. For example, if one day we would succeed in making humans immortal, then everything we had ever thought concerning death and its profundity would become simply laughable [...]. One can surely maintain that this is simply too high a price to pay for the removal of death.[7]

Such thought experiments may seem totally harmless, were it not be for the astonishing fact that a growing number of researchers in artificial intelligence now think it is possible in principle to download consciousness onto a hard drive and in this way bestow virtual immortality onto the individual mind. Hence, if Arendt says of science-fiction literature that "unfortunately nobody yet has paid the attention it deserves,"[8] this means that, in any case, she had then already decided to no longer take for granted the commonly accepted dividing line between reality and fiction and that she was prepared to follow even the most speculative of paths. In fact, in what follows it will be suggested that we ought to read *The Human Condition* as one last great attempt to describe in rigorous phenomenological fashion the three most fundamental human activities as Arendt thought they were on the verge of being swept away by technological change. From this perspective, the book reads as a hardly veiled analysis of the historico-intellectual genesis of technoscience and its possible impact on "those general human capacities which grow out of the human condition and are permanent, that is, which cannot be irretrievably lost so long as the human condition itself is not changed."[9]

---

7 Hannah Arendt, *Denktagebuch 1950-1973*, ed. U. Ludz and I. Nordmann (München/Zürich: Piper, 2003), 591. The translation is mine.

8 Arendt, *The Human Condition*, 2.

9 *Ibid.*, 6.

The three fundamental human capacities (labor, work, and action) correspond to the most general human conditions, "birth and death, natality and mortality."[10] Yet, although the question of a technological induced "removal of death" still features prominently in the prologue – as can be gauged from her poignant remark that man's "wish to escape the human condition, I suspect, also underlies the hope to expand man's life-span far beyond the hundred-year limit"[11] – the remainder of the book contains one long drawn out reflection on the fate of the condition of natality in an increasingly technological driven society. Arendt's remark that natality constitutes "the central category of political, as distinguished from metaphysical thought,"[12] is usually interpreted as implying that she wrote *The Human Condition* out of a dissatisfaction with Heidegger's prioritization of the event of death in his existential analytic of Dasein.[13] This rather negative explanation of why the event of birth holds such a prominent place in her thought, while certainly not wrong, seems to miss the sense of historical urgency which prompted Arendt to write a phenomenology of the active life. If action, the most political of human activities, is rooted in the fact of natality, and natality currently stands exposed to the threat of technological usurpation, then the question of biotechnology, so Arendt in the prologue, "is a political question of the first order."[14] That is to say, it is not just a practical question that professional politicians need to address by enacting legislative measures. It is also, and even more so, a question in which the essence of the political itself is at stake. If human beings are no longer born, but 'made,' then their capacity to act freely may undergo a metamorphosis so great as to become unrecognizable; the capacity, let us not forget, that enables them to make political decisions about such issues as technological development in the first place. There is no question that this kind of threat to the essence of the political is unprecedented. Arendt, however, submits that, when viewed from a broader perspective, the age of biotechnology actually represents but one episode in a much longer long history of the decline of the political. For another important thread that runs through *The Human Condition* is the eclipse of the political as the phenomenological structure informing the capacity to act is perpetually misunderstood from the times of the ancient Greek polis to the present

10 *Ibid.*, 8.

11 *Ibid.*, 2.

12 *Ibid.*, 9.

13 See for example Patricia Bowen-Moore, *Hannah Arendt's Philosophy of Natality* (Basingstoke/London: The Macmillan Press, 1989), 2-5 and Elisabeth Young-Bruehl, *Hannah Arendt. For Love of the World* (New Haven/London: Yale University Press, 1982), 76.

14 *Ibid.*, 3.

day world dominated by the wordless animal laborans. Arendt concludes the book by arguing that not everything is lost though, because, as she suggests in the very last section, "the capacity for action [...] is still with us, although it has become the exclusive prerogative of the scientists."[15] Scientific experimentation is itself an expression of our capacity to introduce unpredictability in the world whose results are not yet knowable. Hence, fully aware that she is writing "at the very moment when [the nature of society] was overcome by the advent of a new and yet unknown age,"[16] she does not venture to make a prediction about how the human capacity to act will fare in the dawning age of pervasive biotechnology.

Although, arguably, the biotechnological threshold had already been crossed with the introduction of IVF in 1978, something which Arendt discussed as early as 1958 – "creating life in the test tube"[17] –, the pertinence of her thought for reflecting on biotechnological issues was only discovered quite recently. As we have seen, in *The Future of Human Nature* Habermas strongly opposes non-therapeutic genetic intervention and even suggests that it should be legally banned on the basis of a "right to a genetic inheritance immune from artificial intervention"[18] because it would radically infringe on the genetically programmed person's capacity to act freely.[19] For a person to know herself as the irreducible origin of her own free actions, Habermas explains, it seems necessary that she experiences her actions with reference to an origin that exceeds her and somebody else's control. Such a condition, he suggests, is what Arendt attempted to articulate through the concept of natality: "In acting, human beings feel free to begin something new because birth itself, as a divide between nature and culture, marks a new beginning."[20] In order to experience oneself as the unique origin of authentic actions, the distinction between "what is manufactured and what has come by nature,"[21] between what has been brought into being by *tekhnē* and what has come into be-

15 *Ibid.*, 323.

16 *Ibid.*, 6.

17 *Ibid.*, 2.

18 Habermas, *The Future of Human Nature*, 27.

19 Michael Sandel has expressed a similar concern about the corrosive effect of biogenetic engineering on one's ethical freedom to be the undivided author of one's own life. In his argument against genetic enhancement he also avails himself of Arendt's concept of natality and argues that biotechnological intervention exhibits "a stance of mastery and domination that fails to appreciate the gifted character of human powers and achievements, and misses the part of freedom that consists in a persisting negotiation with the given" (Sandel, *The Case Against Perfection*, 83).

20 Habermas, *The Future of Human Nature*, 59.

21 *Ibid.*, 46.

ing by *physis*, seems categorical. It is only because we are aware that an essential aspect of our identity was already fixed at the time of our birth, that we are capable of experiencing ourselves as the initiators of actions for which we are uniquely responsible:

> What is suggested by [the concept of natality] is, I believe, the onset, with birth, of a differentiation between the socialization fate of a person and the natural fate of her organism. It is only by referring to this difference between nature and culture, between beginnings not at our own disposal, and the plasticity of historical practices that the acting subject may proceed to the self-ascriptions without which he could not perceive himself as the initiator of his actions and aspirations [...]. The fact that this natural fate, this past before our past, so to speak, is not at our human disposal seems to be essential for our awareness of freedom.[22]

Habermas acknowledges that from Arendt's description of the condition of natality it does not necessarily follow that a genetically programmed person will no longer be able to experience her actions as constitutive of new beginnings, but he gives an additional reason why, nonetheless, this will very likely be the case. Unlike the situation of someone who finds herself confronted with the expectations and stimulations of her parents to act upon her natural talents given at birth, a programmed person can only interpret, but not revise or undo her parents' choice to purposefully intervene in her genetic make-up. In other words, genetic intervention will establish a paternalistic, asymmetrical relationship that – as he had already argued in an earlier essay[23] – even surpasses slavery in cruelty, for it will never be possible to emancipate oneself from this inferior, programmed position.

Natality Between Necessity and Freedom

Reflecting on Marx's seemingly incompatible statements that labor is the most human of man's activities and that the abolition of labor will inaugurate the realm of freedom, Arendt writes:

> Such fundamental and flagrant contradictions rarely occur in second-rate writers, in whom they can be discounted. In the work of great authors they lead into the very center of their work and are the most important clue to a true understanding of their problems and new insights.[24]

---

22 *Ibid.*, 59-60.

23 See Jürgen Habermas, 'An Argument against Human Cloning. Three Replies,' in *The Postnational Constellation. Political Essays*, tr. and ed. M. Pensky (Malden, Mass.: Polity Press, 2007), 163-172. Originally published in Süddeutsche Zeitung, January 17-18, 1998.

24 Hannah Arendt, *Between Past and Future. Eight Exercises in Political Thought* (New York: Penguin, 2006), 24.

Thus, if it is true that the real originality of great authors reveals itself especially in the contradictions in their work, then one cannot claim to have really understood the thought of Arendt, nor claim that it contains an argument against human genetic modification unless the meaning and sense of the concept of natality has been completely unraveled. Natality, Arendt explains in *The Human Condition*, names the fact that "[b]ecause they are initium, newcomers and beginners by virtue of birth, men take initiative, are prompted into action."[25] But is it not paradoxical for Arendt either to argue that natality constitutes the "central category of political thought"[26] if she also maintains that it is through the activity of labor, not action, that men deal with the biological processes forced upon them by necessity? In other words, how can Arendt both claim that political freedom is rooted in such a plain biological phenomenon as parturition and that everything related to "the life of the individual and the survival of the species" is a "non-political, household affair by definition"[27]?

Indeed, it is striking that Arendt is not very consistent in her description of the nature of and the relation between the activities pertaining to the realms of necessity and freedom respectively. In the first part of *The Human Condition*, she appears to accept the ancient Greek view and makes a clear-cut separation between the private and the public realm, *oikos* and *polis*, even to the point of endorsing the Greek idea that it is legitimate to use violence in the private sphere on condition that it serves the purpose of liberating "oneself from the necessity of life for the freedom of world."[28] Arendt's "phenomenological essentialism,"[29] as Seyla Benhabib has called her one-to-one coupling of human activities and worldly locations, seems to lead her to accept the Greek view that we are all born as unequal, needy creatures irredeemably tied to the necessities of the life process, but that we are nevertheless intermittently capable of raising ourselves out of this slavish condition by erecting an artificial public realm in which our biological wants and needs can be temporarily suppressed:

> The distinctive trait of the household sphere was that in it men lived together because they were driven by their wants and needs. [...] The realm of the polis, on the contrary, was the sphere of freedom, and if there was a relationship between these two spheres, it was a matter of course that the mastering of the necessities of life in the household was the condition for the freedom in the polis.[30]

25 Arendt, *The Human Condition*, 177.

26 *Ibid.*, 9.

27 *Ibid.*, 29.

28 *Ibid.*, 31.

29 Seyla Benhabib, *The Reluctant Modernism of Hannah Arendt* (Oxford: Rowan & Littlefield, 2003), 123.

30 Arendt, *The Human Condition*, 30-31.

From remarks like these it has often been concluded that Arendt suffered from an ardent "polis envy," in the sense that she would have posited the socio-political structure of the Greek polis as a normative ideal against which modern society, with its conspicuous "unnatural growth of the natural,"[31] can be critically measured.[32] In other words, her virulent hostility toward what she calls "the rise of the social," the encroachment of the activities having to do with life, labor, and reproduction on the political realm, is seen by many as the result of the fact that she made her idiosyncratic understanding of the Greek polis life the standard against which modern society is assessed and, obviously, found wanting.

However, there are also passages in which she seems to argue the exact opposite, namely, that there is nothing that more clearly reveals the anti-political thrust of the Western tradition than its deep-seated drive to violate biological life. This is, for example, the case at the end of the chapter of *The Origins of Totalitarianism* entitled 'The Decline of the Nation-State and the End of the Rights of Man.' Describing the fate of the millions of displaced and stateless persons in interbellum Europe, Arendt argues that these people were deprived of their human rights the moment they most needed them. Both the French and the American proclamations of the Rights of Man state that each human being is endowed with a set of fundamental rights that are "self-evident" or "given with birth." Yet, the very moment one lost one's political status as a citizen of the nation-state, it turned out that one could not appeal to these so-called "inalienable rights." Quite the reverse, more than often these stateless people found themselves reduced to "the abstract nakedness of being human and nothing but human,"[33] the extreme case being the camp inmate reduced to nothing more than "a specimen of the animal-species man."[34] Totalitarianism can be said to have brought this violent, anti-political drive to its most extreme conclusion, but in fact it has determined Western politics since its inception. "Ever since the Greeks," Arendt explains at the very end of the chapter, "we have known that highly developed political life breeds [...] a deep resentment against the disturbing miracle contained in the fact that

---

31 *Ibid.*, 47.

32 See for example Bernard Flynn, *Political Philosophy at the Closure of Metaphysics* (Atlantic Highlands: Humanities Press, 1992); Maurizio Passerin d'Entrèves, *The Political Philosophy of Hannah Arendt* (London: Routledge, 2003); Hanna Pitkin, *The Attack of the Blob. Hannah Arendt's Concept of the Social* (Chicago: The University of Chicago Press, 2000) and Richard Wolin, *Heidegger's Children. Hannah Arendt, Karl Löwith, Hans Jonas, and Herbert Marcuse* (Princeton, NJ: Princeton University Press, 2001).

33 Arendt, *The Origins of Totalitarianism*, 377.

34 *Ibid.*, 589.

each of us is made as he is – single, unique, individual."[35] The dark background formed by what is biologically given to us at birth "breaks into the political scene as the alien which in its all too obvious difference reminds us of the limitations of human activity" and "indicates those realms in which man cannot change and cannot act and in which, therefore, he has a distinct tendency to destroy."[36] Contrary to what she will argue ten years later in *The Human Condition*, then, Arendt concludes that such a suppression of biological life has to be avoided at all costs:

> The human being who has lost his place in a community [...], is left with those qualities which usually can become articulate only in the sphere of private life and must remain unqualified, mere existence in all matters of public concern. This mere existence, that is, all that which is mysteriously given us by birth and which includes the shape of our bodies and the talents of our minds, can be adequately dealt with only by the unpredictable hazards of friendship and sympathy, or by the great and incalculable grace of love, which says with Augustine, "*Volo ut sis* [I want you to be]," without being able to give any particular reason for such supreme and unsurpassable affirmation.[37]

It obviously goes without saying that such a "basic gratitude for everything that is as it is," for "things that are *physei* and not *nomō*,"[38] is diametrically opposed to what Arendt, apparently approvingly, described as the ancient Greek hierarchy between the biological and politico-linguistic dimensions of man.

But again, how then should we conceive the relation between birth and the freedom to act that grounds Arendt's concept of natality? And how will this condition fare in the dawning age of biotechnology? Does biological birth really condition the capacity to act, in which case Habermas seems to be fully justified in invoking the concept of natality to oppose the use of technologies that bring qualities that are given by birth within the reach of planned technical intervention? Or are they antithetical notions, as Arendt also seems to argue, in which case the concept of natality would rather turn into a weapon in the hands of those, like the liberal eugenicists, who believe that our present physical and psychical constitution is an obstacle that impedes the realization of our full potentialities and who therefore belief that that prospective parents should be given as much freedom as possible to decide which genetic traits their children will or will not have upon birth?

---

35 *Ibid.*, 382.

36 *Ibid.*, 383.

37 *Ibid.*, 382.

38 Hannah Arendt, *The Jew as Pariah* (New York: Grove Press, 1978), 246.

The Symbolic Reduction of the Event of Parturition

Many of Arendt's commentators have circumvented the above-mentioned contradiction in her description of the relation between birth and action by assuming that she made a sharp distinction between biological birth and politico-linguistic birth, *zoe* and *bios*, and that for her only the latter kind of birth genuinely expresses the fact of natality.[39] According to Ronald Beiner, for example, Arendt "conforms to a strict dualism of Nature and Freedom"[40] and he even goes as far as claiming that Arendt's concept of natality forms the core of what he calls her "anti-naturalistic philosophy."[41] Such an understanding of natality fits nicely into the standard interpretation of *The Human Condition*, which holds that it obviously could not have been Arendt's intention to root political freedom in the biological event of parturition, because all evidence points to the conclusion that Arendt bemoaned the victory of the animal laborans by unfavorably comparing its mode of existence to what the Greeks understood as a genuine *bios politikos*. Moreover, these commentators point out, the capacity to begin something new presupposes a form of temporality that is fundamentally at odds with the temporality that humans experience in their biological processes, which are, as Arendt explains, "part of the cyclical movement of nature and therefore endlessly repetitive."[42] Contrary to the eternal circularity of natural processes, the processes that the human sets into motion through action have a definite beginning in time. The fact that Arendt invokes the event of the birth of a child to characterize the emergence of the new inherent in political action should, according to them, be understood from the perspective of the strategic value that it possesses. For it provides an age dominated by the worldless animal laborans with an example of the one natural event that discloses the same mode of temporality that structures worldly action. Put otherwise, on this reading, politico-linguistic birth should be seen as the supreme actualization of the potentiality for beginning something new that biological birth only expresses *symbolically*.

The assumption that biological birth stands merely for the human's capacity to initiate new beginnings would then find further confirmation

39 See for example Margaret Canovan, *The Political Thought of Hannah Arendt* (London: Methuen, 1974); Ronald Beiner, 'Action, Natality and Citizenship: Hannah Arendt's Concept of Freedom,' in *Conceptions of Liberty in Political Philosophy*, ed. Z. Pelczynski and J. Gray (London: Athlone Press, 1984) and Margarete Durst (2003), 'Birth and Natality in Hannah Arendt,' *Analecta Husserliana* Vol. 79, 777-797.

40 Beiner, *Action, Natality and Citizenship*, 361.

41 *Ibid.*, 358.

42 Arendt, *The Human Condition*, 98.

in Arendt's 1929 doctoral dissertation *Love and Saint Augustine* in which she explores the close connection between human creatureliness and freedom. The central aim of this early study was to question the possibility of grounding the precept "Love thy neighbor as thyself" in the Christian notion of love of God. In the first chapter of the book, Arendt shows that Augustine's notion of love as craving is not suitable for this purpose. Since it presupposes that the "highest good" (i.e. the "happy life") can only be attained in an absolute future which lies beyond worldly life and death, this kind of love demands complete self-denial and forsakenness of the human world and hence, so Arendt, "makes the central Christian demand to love one's neighbor as oneself well-nigh impossible."[43] In the chapter entitled 'Creator and Creature,' Arendt shows, however, that Augustine's writings contain yet another notion of love which is not primarily oriented toward an absolute future, but toward an absolute past: "When happiness is projected into the absolute future, it is guaranteed by a kind of absolute past, since the knowledge of it, which is present in us, cannot possibly be explained by any experience in this world."[44] In order to love happiness, one must already know what happiness is, and this knowledge "is given in pure consciousness prior to all experience."[45] Hence, Arendt concludes, the only way to reach the idea of happiness is through remembrance: "[S]ince recollection presents a knowledge that necessarily lies before every specific past, it is also truly directed toward a transcendent and transmundane past – that is, toward the origin of human existence as such."[46] This other notion of love, then, is a love of God that can only be actualized through a return, in recollection, to the One who created man:

> The creature in its createdness derives its sense of meaning from a source that precedes its creation, that is, from the Maker who made it [...]. The fact that man has not made himself but was created implies that the meaningfulness of human existence both lies outside itself and antedates it.[47]

Arendt confirmed the assumption that her early work on Augustine was the zero-point of her philosophy of natality by adding a passage in the 1964-5 revised English edition in which she explicitly calls the relation between human creatureliness and meaningfulness an expression of

43 Hannah Arendt, *Love and Saint Augustine* (Chicago: The University of Chicago Press, 1996), 30.

44 *Ibid.*, 47.

45 *Ibid.*

46 *Ibid.*, 48.

47 *Ibid.*, 50.

the fact of natlity.[48] But the mere fact that it was also possible for her to develop a theological understanding of natality would then be yet another indication that the biological understanding of natality that she develops in The Human Condition is but one historico-paradigmatic figure of the capacity to initiate new beginnings that finds its most basic experience in politico-linguistic action.

Yet, if Arendt really intended to say that there is merely a structural or symbolic analogy between the unexpected arrival of a new human being through birth and the emergence of the new through worldly action, just as there is an analogy, as she argues elsewhere, between divine creation *ex nihilo* or the unexpected birth of Jesus of Nazareth and the "miracle of action,"[49] then why does she, nonetheless, claim that in this politico-linguistic birth "we confirm and take upon ourselves the naked fact of our original physical appearance"[50]? Moreover, it is also highly unlikely that the ancient Greek separation between the realms of necessity and freedom, *oikos* and *polis*, and between the modes of existence appropriate to each, *zoe* and *bios*, underpins Arendt's concept of natality, if only because she repeatedly insists that it names a concept "which Greek antiquity ignored altogether."[51]

In a number of recent publications on Arendt's concept of natality it has, therefore, been forcefully argued that the reference to biological birth was not merely a metaphorical gesture, but that she really wanted to say that the human capacity to act is literally conditioned by the bare biological fact of being born. Neither, however, was it an attempt to ground political action in an immutable human nature, as also has been suggested.[52] Peg Birmingham, for example, has argued that Heidegger's notion of solicitude (*Fürsorge*) could help us to understand that the newborn's entrance into the world is never simply a physical event, but always already also a politico-linguistic event: "Linguistic natality cannot be laid over physical natality, and this suggests that both births are inseparable and always found together."[53] It is thus not the case, she explains, that for

48 *Ibid.*, 52-53.

49 "It is this faith in and hope for the world that found perhaps its most glorious and most succinct expression in the few words with which the Gospels announced their 'glad tidings': 'A child has been born unto us'" (Arendt, *The Human Condition*, 247).

50 *Ibid.*, 176.

51 *Ibid.*, 247.

52 See for example Martin Jay, 'The Political Existentialism of Hannah Arendt,' in *Hannah Arendt: Critical Assessments of Leading Political Philosophers*, Vol. 3, ed. G. Williams (New York: Routledge, 2006), 191-213.

53 Peg Birmingham, *Hannah Arendt and Human Rights. The Predicament of Common Responsibility* (Bloomington: Indiana University Press, 2006), 25.

Arendt the newborn is simply a specimen of the animal-species man who only actualizes her potentiality to become a unique self when she inserts herself into the public world through words and deeds. "Naked facticity," Birmingham writes, "is always already the site of language,"[54] namely, in the sense that the newborn is always immediately exposed to the welcoming address of the other.

Anne O'Byrne agrees with Birmingham that biological birth and politico-linguistic birth are so deeply implicated as to be inseparable, but she retorts that by identifying their moment of convergence with the naming of the newborn, Birmingham still sets too great a distance between the biological and the linguistic components of the event of natality. The problem with this understanding of natality, O'Byrne explains, is that it expels the intimate relationship between the maternal body and the fetus to the presocial and prelinguistic sphere and that it privileges the paternal moment in the event of natality. O'Byrne's notion of natality's syncopated temporality offers a way out of this impasse. By this mode of temporality she means "a mode of being in time that can grasp itself only belatedly."[55] If the moment of my physical birth constitutes a past that was never present to me, because I was not 'there' to experience it, but at a later moment in time still turns out to have been my birth, then the temporal structure of this event also determines our politico-linguistic birth. The outcome or meaning of one's actions also only reveals itself to the backward glance of the storyteller or the historian, never to the actor himself – the meaning of one's deed only reveals itself after the event. Following O'Byrne, then, it can be argued that the intimate connection between both kinds of births shows itself by the fact that our politico-linguistic birth always arrives too late, in the sense that our biological birth has always already 'happened' to us. No one was present at her own birth. This event constitutes an absolute past which necessarily remains outside our field of experience. Yet, one could wonder whether even Anne O'Byrne's articulation of the co-implication of biological birth and politico-linguistic birth does not still leave too great a gap between both events. After all, both the ability to reflect on one's moment of birth and the ability to understand the stories that others tell about this event develop only gradually over time, while Arendt clearly suggests that both events always occur together. Arendt is quite clear about the fact that the mere event of being born alone predisposes one to act freely. Therefore the task remains to think the co-implication of biological birth and politico-linguistic birth in still more radical fashion.

---

54 *Ibid.*, 29.

55 O'Byrne, Natality *and* Finitude, 95.

The Prematurity of Natal Life

When, in a crucial yet largely neglected footnote in *The Human Condition*, Arendt states emphatically that her description of the condition of natality is "supported by recent findings in psychology and biology"[56] and refers to the German anthropologist Arnold Gehlen's book *Der Mensch* [*Man*] as her main source of inspiration, she is most likely referring to what in the field of evolutionary biology is called the phenomenon of neoteny or fetalization. In the first chapter of his book, Gehlen discusses the Dutch anatomist Louis Bolk's essay 'Das Problem der Menschwerdung' [*The Problem of the Origin of Man*, 1926].[57] Bolk wrote this essay as a challenge to Ernst Haeckel's then still widely supported theory of recapitulation. He argues that Haeckel's theory of evolution can indeed explain the mechanisms that determine the ontogenetic development of animal species, but that it cannot provide an answer to the much more pressing riddle of human ontogentic development. Bolk departs from the often quoted but never quite satisfactorily explained observation that adult humans strongly resemble juvenile pongids, but that this phenotypic likeness gradually disappears during the pongid's ontogenetic maturation. Moreover, in contrast to humans, in pongids there is a strong negative allometry of the brain and a strong positive allometry of the jaws. According to Bolk, these phenomena cannot be explained by the thesis that in ontogenetic development humans go through the different stages that determined the phylogenetic evolution of their direct ancestors. His alternative theory holds that, in contrast to animals, humans evolved by retaining a number of juvenile and even fetal features of their direct ancestors throughout ontogenesis. Hence, whereas in the ontogenetic development of non-human primates bodily traits such as a flat face, a reduction of body hair, and high relative brain weight represent only temporary features, in humans they have evolved to become permanent features of their physical constitution.

There is no immediate evidence that Arendt ever engaged herself in a thorough study of Bolk's theory of fetalization. Nonetheless, it is quite plausible that one implication in particular could have drawn her attention while reading Gehlen's concise rendering of it. Bolk further specified that the retention of fetal characteristics in humans can be explained by the occurrence of a general retardation of human ontogenetic development, itself caused by an alteration of the endocrine system. In other words, it is because maturation is delayed in humans that fetal growth rates are

---

56 *Ibid.*, 178.

57 Arnold Gehlen, *Man: His Nature and Place in the World*, tr. C. McMillan and K. Pillemen (New York: Columbia University Press, 1988), 93-109.

prolonged and fetal features stabilized. Retardation explains, for example, why humans live much longer than other primates and mammals of comparable body size. A more interesting implication, however, is that if humans were to attain the same level of ontogenetic development as other primates at their time of birth, they would actually need a gestation period of twenty-one months instead of the nine months now. In a sense, the Swiss biologist Adolf Portmann noted, one could therefore say that humans spend their first year as "extrauterine embryos."[58] The main reason for this acceleration of time of birth is that even at this stage of ontogenetic development the human brain continues to grow at fetal rates. Humans achieve only twenty-three percent of their full brain capacity at term, whereas the brains of other mammals are at that time already fully formed. But if this growth of the brain would have to take place in the uterus, then it would be physically impossible for a woman to give birth.

Gehlen believed he had found in Bolk's theory of fetalization hard scientific evidence for his conception of man as a "deficient being" [*Mängelwesen*] and it seems that Arendt's reading of the former inspired her to develop her concept of natality on the basis of a similar anthropological theory. Unlike animals, Gehelen argues, humans are born without any well-developed instincts and without specialized organs and are thus singularly unfit for survival. He therefore rejects the standard interpretation of the evolutionary theory of the origin of man. Humans are not so much superior to other animal species, but are, on the contrary, vastly inferior to them: "One envisions man fictitiously as animal only to discover that he makes an imperfect and indeed impossible animal."[59] Thus, when Arendt states in the mentioned footnote to Gehlen that the scientific theories he discusses allow for the conclusion that action and speech are a "'biological necessity,' that is, necessary for a biologically weak and ill-fitted organism such as man,"[60] she clearly inscribes herself in a peculiarly German tradition – initiated by Herder and brought into prominence by Nietzsche – that understands humans as 'indeterminate' or 'deprived' animals. Whereas newborn animals are almost immediately capable of generating appropriate reactions to the stimuli that emerge out of their environment, humans are extremely ill-adapted to the environment into which they are thrown. Born prematurely and thus deprived of any particular biological quality, no such spontaneous attunement between human organism and environment takes place. Because of their premature birth,

58 Adolf Portmann (1941), 'Die Tragzeiten der Primaten und die Dauer der Schwangerschaft beim Menschen: ein Problem der vergleichen Biologie,' *Revue Suisse de Zoologie*, Vol. 48, 511-518. Gehlen refers to this article in the first chapter of *Man*.

59 Gehlen, *Man*, 13.

60 Arendt, *The Human Condition*, 177.

humans enter the world helpless and needy and thus in desperate need of protection and care by the social group. According to Portmann, one can even argue that the social group assumes the task of an "external uterus." It is in this sense that the theory of fetalization allows us to understand more clearly why Arendt can argue that natality is an inextricable biological and politico-linguistic event, for as premature creatures humans are biologically conditioned to engage in politico-linguistic action. As Gehlen puts it, "a being with such a physical constitution is viable only as an acting being."[61] Thus, when Arendt writes that "a life without speech and without action [...] has ceased to be a human life,"[62] she clearly remains within the confines of Gehlen's anthropological theory.

This makes it understandable why Arendt can argue that the concept of natality articulates the idea that human freedom is conditioned by the biological fact of being born without necessarily contradicting herself. Natality not only articulates the fact that our politico-linguistic birth always arrives too late, in Anne O'Brien's sense that our biological birth constitutes a past that was never present to us. The concept of natality also articulates the fact that our biological birth always arrives too early, namely, in the sense that our premature birth releases us from the fate of being compelled to follow a biogenetically predetermined course of life. This is also probably one of the reasons why Arendt rejects the notion of human nature and opts to speak about human conditions instead.[63] Natality, as a condition of human existence, does not determine human beings absolutely because the biological traits they are born with never solidify into a set of fixed properties but remain in a deficient state. From a biological perspective, we were never ready to enter the world in the first place. But it is precisely this 'unpreparedness' or 'prematurity' that gives us to the possibility of initiating radically new beginnings. Natality, as we can see now, is not a purely biological concept, nor a purely politico-linguistic one. It names an event which breaks out of the eternal circle of nature, "where no beginning and no end exist,"[64] but which nonetheless remains ineluctably tied to the biological condition from which it emerged.

The freedom of natal beings consists in an "persisting negotiation with the given,"[65] to use Michael Sandel's words, but the foregoing suggests

---

61 Gehlen, *Man*, 16.

62 Arendt, *The Human Condition*, 176.

63 "[T]he human condition is not the same as human nature, and the sum total of human activities and capabilities which correspond to the human condition does not constitute anything like human nature" (*ibid.*, 9-10).

64 *Ibid.*, 96.

65 Sandel, *The Case Against Perfection*, 83.

that actually very little has been given to them at birth. Deprived of an immutable 'natural essence' and not tailored to any particular environment, humans are biologically conditioned to actively create their own conditions of existence:

> In addition to the conditions under which life is given to man on earth, and partly out of them, men constantly create their own self-made conditions, which, their human origin and their variability notwithstanding, possess the same conditioning power as natural things. Whatever touches or enters into a sustained relationship with human life immediately assumes the character of a condition of human existence.[66]

Furthermore, Arendt makes it utterly clear that the condition of natality not only prompts them to engage in politico-linguistic action. Natality also prompts them to engage in technological innovation:

> [I]f the human condition consists in man's being a conditioned being for whom everything, given or man-made, immediately becomes a condition of his further existence, then man "adjusted" himself to an environment of machines the moment he designed them. They certainly have become as inalienable a condition of our existence as tools and implements were in all previous ages.[67]

Contrary to what Habermas and Sandel believe, then, Arendt does not believe that biotechnological intervention will irrevocably infringe on the human capacity to act freely. In her view, technology is not an external force that threatens to violate a pure, natural origin, but constitutes an irreducible condition of human existence. Technology could even be said to constitute natality's necessary 'other,' namely, in the sense that it is the originary supplement of the human's faulty biological constitution. If humans are born deprived of any fixed set of biological characteristics, if they are born 'unfinished,' as it were, then simply opposing oneself against genetic intervention becomes an impossible position to sustain. Of course, this does not mean that genetic intervention is entirely unproblematic and thus exempted from critical examination. But it does seem to imply that it is impossible to determine in advance what is 'natural' and what is 'unnatural' to the existence of natal beings. Abiding in the gap between past and future, natal beings are destined to invent, to explore, and to go where nobody has ever went before, imbued as they are with the promise of always being capable to make a fresh start.

---

66 Arendt, *The Human Condition*, 9.

67 *Ibid.*, 147.

CHAPTER FIVE

Prosthetic Life[1]

> [A]t stake here will be the attempt to think, instead of the birth of the human *qua* entity related to its end, rather its invention or even its embryonic fabrication or conception, and to attempt this independently of all anthropologism, even if this would mean considering with the utmost seriousness this other question: "And if we already were no longer human?"[2]

The Forgetting of Epimetheus

Bernard Stiegler's ongoing philosophical project *Technics and Time*[3] is one of the most original and promising efforts in contemporary continental philosophy to rethink the relationship between the human and technology.[4] Drawing on the writings of Leroi-Gourhan, Husserl, Heidegger, Derrida, and Simondon, Stiegler seeks to argue that "anthropogenesis corresponds point by point to a technogenesis,"[5] in the sense that human beings and technical artifacts have always been involved in a mutually constitutive relationship. In the first volume of *Technics and Time*, 'The Fault of Epimetheus,' Stiegler sets up his argument around the emblematic figure of Epimetheus, Prometheus' slow brother whose role in the infamous myth, just as that of technics, is usually forgotten by the philosophical tradition.[6] Since this story contains in broad outlines the central

1 Parts of this chapter were published as Nathan Van Camp (2009), 'Stiegler, Habermas, and the techno-logical condition of man,' *Journal for Cultural Research* 13 (2), 125-141 and Nathan Van Camp (2011), 'Negotiating the Anthropological Limit: Derrida, Stiegler, and the Question of the Animal,' *Between the Species*, 14 (1), 57-80.

2 Bernard Stiegler, *Technics and Time 1: The Fault of Epimetheus*, trans. R. Beardsworth and G. Collins (Stanford: Stanford University Press, 1998), 135-136.

3 *The Fault of Epimetheus*, the first volume of this series, appeared in 1994, followed by *Disorientation* in 1996, and *Cinematic Time and the Question of Malaise* in 2001. Stiegler has planned to write two more volumes in the near future.

4 While perhaps overstating it, Richard Beardsworth even goes as far as claiming that "it is a work the importance and effects of which can be compared, in the continental tradition at least, with Heidegger's *Being and Time* and Derrida's *Of Grammatology*." See Ricard Beardsworth (1995), 'From a genealogy of matter to a politics of memory: Stiegler's thinking of technics,' *Tekhnema*, 2, 2.

5 Stiegler, *The Fault of Epimetheus*, 45.

6 See *ibid.*, 185-203.

features of Stiegler's theory, it is worthwhile to summarize it briefly. In Protagoras' version of the story as narrated by Plato, Epimetheus convinces his brother Prometheus to entrust him with the task of distributing the qualities or powers among the living creatures that was commended by Zeus. The Titan performs his task by allotting each creature one specific quality making sure that there is a just equilibrium between the species. But when it was time for the humans to receive their due, Epimetheus suddenly realizes that he had already handed out all available qualities. When Prometheus arrives at the scene to inspect his brother's work, he notices that the humans are still naked and without means of defense. Because he is worried that their premature constitution could lead to the total destruction of mankind, Prometheus decides to make up for Epimetheus' fault and goes out to steal the technical skills and fire from the gods to compensate humans for their lack of qualities. What Stiegler wants to stress in his reading of this myth is the fundamental undefinability of the human or what he calls the human's "default of origin." Hence, technics is not proper to the human either, but merely an external prosthesis that is only adopted afterwards, by default, to compensate for an originary lack of origin. According to Stiegler, the figure of Prometheus, who is usually interpreted as the symbol of the technical hubris of man, can therefore only be understood in connection to that of his brother Epimetheus. The fact that the latter's role is usually forgotten by the philosophical tradition is just as remarkable as it is understandable, given that Epimetheus is himself essentially a figure of forgetting. What this myth makes clear is that the human being is the forgetful one, the one who, in Stiegler's terms, suffers from "retentional finitude" and who therefore needs technical prostheses to supplement his limited capacity for memorization.

The meaning of this mythical narrative can also be explained in bioanthropological terms. According to Stiegler, memory can be articulated on three different levels. First, there is genetic memory, which is common to all living beings and which remains strictly internal to the organism. Second, there is epigenetic memory, which consists of individually acquired experiences that are stored in the central nervous system of the organism. In contrast to genetic memory, epigenetic memory provides the organism with a certain degree of latitude, which explains why most species of animals can be conditioned to behave in certain ways. The main difference between genetic memory and epigenetic memory is that while the former is automatically transferred from one generation to the other, the latter is irremediably lost for the species with its carrier's death. Put differently, these two levels of memory don't communicate with each other. Stiegler claims, however, that with the 'birth' of the human there emerged a third kind of memory – epiphylogenetic or technical memory –, which allows the human to overcome the evolutionary impossibility to transmit epigenetic memory to the next generations. The reason for this is that, in contrast to the animal, the human being is capable of exteriorizing his epigenetic memory onto material supports (from flint tools and wax tablets to combustion engines and the TCP/IP protocol) that can be preserved over time.

What has so far not been sufficiently emphasized in the recent literature is that the starting point of Stiegler's theory of technics was the problem of genetic intervention. The fact that this has previously gone unrecognized is hardly surprising, given that Stiegler himself confined his remarks on this subject to a short essay dedicated to Derrida which only appeared in French and, moreover, predates the publication of the *The Fault of Epimetheus*. Nevertheless, it could be argued that not only can one not fully comprehend his theory of technics without acknowledging this initial starting point, but also, as we will argue further on, that his theory of technics has enormous consequences for the mode in which the debate on genetic intervention is currently conducted. In 'Quand faire c'est dire'[7] ['When to do is to say'], Stiegler notes that genetic intervention is "at the same time *radically new* and *absolutely ancient*."[8] In order to understand this claim, we have to recall that, in chapter one, we showed that with the emergence of genetic technologies the human species has entered an entirely new phase in evolution. Our point was that if it is possible to deliberately introduce changes in the human genome, then we are currently witnessing the suspension of the Central Dogma of Molecular Biology upon which the whole of Neo-Darwinism is constructed. Remember that this Dogma holds that the genetic program does not take lessons from experience, in the sense that whatever knowledge or skills an individual organism might acquire during its life, all this is forever lost for the species when the individual organism dies. The radical novelty is, then, that insofar as genetic technologies allow to deliberately introduce changes in the human genome, it appears that the genetic program can receive lessons from experience after all. Yet, Stiegler argues, this is perhaps not as revolutionary as might seem at first sight, because it could very well turn out "that the 'laws of evolution' have already been suspended for quite some time now – at least since the invention of the human, that is, since the invention of technics."[9] This claim is far more radical than what has hitherto been assumed. What Stiegler means to say is not simply that the human is the only living being that is capable of preserving his memories over time by grafting them onto technical objects. He also intends to say

7 Bernard Stiegler, 'Quand faire, c'est dire. De la technique comme différance de toute frontière,' in *Le passage des frontières. Autour du travail de Jacques Derrida*, ed. Colloque de Cérisy (Paris: Galilée, 1994), 271-283.

8 *Ibid.*, 272; Translation mine.

9 *Ibid.*; Translation mine.

that one particular species of living beings managed to free itself from the pressure of genetic selection and to become 'human' by entering into a mutually constitutive relationship with technics.

In this chapter we will attempt to clarify in more detail Stiegler's claim that the human is co-constitutive with technics. In order to do this, we will start by examining Derrida's work on the anthropological difference. The reason why it is necessary to devote part of our attention to the question of the human/animal distinction in Derrida is that, as we will show in much more detail below, Stiegler's theory of technics is to a large extent the outcome of a productive dialogue with Derrida on precisely this question. Moreover, it will be shown that much of this discussion centers on the interpretation of the work of the French paleontologist André Leroi-Gourhan. It is on the basis of the latter's work that Stiegler will argue that the human is invented by technics as much as technics by the human. Subsequently, we will use this insight to develop a critical reading of Habermas's theory of technology. It will be shown that Habermas's view that technology poses a threat to human nature is largely due to his mistaken belief that technology is and must remain simply a means to fulfill a human purpose. To conclude, we will suggest that Stiegler's theory of originary technicity urges us to rearticulate the debate on genetic technologies in different terms.

Deconstructing the Anthropological Difference

In a collection of lectures published as *The Animal That Therefore I Am*,[10] Jacques Derrida elaborates a critique of "the gesture [that] seems to constitute philosophy as such,"[11] namely the positing of a single and indivisible dividing line separating the animal from the human. For Derrida, the question of the animal is therefore not one more pressing question among many others, but probably "the most important and decisive question,"[12] a question which he says to "have addressed a thousand times, either directly or obliquely, by means of readings of all the philosophers I have taken an interest in [...]."[13] Hence, what for Derrida secretly unites such diverse thinkers as Descartes, Kant, Levinas, Lacan and Heidegger is that they simply oppose the *Animal* to the *Human* by denying all animals certain qualities they deem to be proper to man alone. To counter this tradition, Derrida will, however, not give back to the

10 Jacques Derrida, *The Animal That Therefore I Am*, trans. D. Wills (New York: Fordham University Press, 2008).

11 *Ibid.*, 40.

12 *Ibid.*, 34.

13 *Ibid.*

animal what man has always deprived it of – a gesture that risks introducing biological continuism and all the dangerous implications thereof –, but argue that all living beings, including the human being, suffer a radical powerlessness.

For many a reader of Derrida, his statement about the central importance of the question of the animal in his work will very likely come very unexpectedly. Indeed, at first sight this claim seems to be in flagrant contradiction with the simple fact that Derrida has not addressed this question as a specific subject of inquiry prior to the publication of *The Animal That Therefore I Am*. To be sure, there has never been a lack of animal figures in Derrida's texts, but the strategic functions these animal figures fulfill are usually highly context-dependent and text-specific and certainly do not set the stage for a more profound investigation of the larger stakes of this issue. Yet, there nevertheless seems to be every reason to take this claim seriously. As if he wants to convince his readers that his is not merely using a hyperbole when making this claim, Derrida bluntly declares in this text that "logocentrism is first of all a thesis concerning the animal."[14] Derrida's versatility as a thinker renders it almost impossible to detect an overarching theme in his writings, but it would certainly not be inaccurate to state that his critique of logocentrism – the claim that there is a substantial self predating and enduring its interactions with the world – motivates nearly all of his readings of texts of the Western philosophical tradition. Looking back at his early writings such as *Of Grammatology* (1974) and *Writing and Difference* (1978) from this perspective, one could indeed argue that the "quasi-transcendental" concepts he elaborated in these works, such as the trace, the *grammē* and archi-writing, can also be read as potentially holding a strong critique of the rigid human/animal distinction bequeathed by the tradition. But even if we grant him the benefit of the doubt and assume that he did not merely retrospectively project a more recent sensitivity onto his early work, this issue was surely not at the center of Derrida's attention at that time. It was therefore generally assumed, and even more so following the so-called ethical turn of deconstruction marked by the growing influence of Levinas's work on Derrida, that his critique of logocentrism was mainly intended to call attention to the many forms of human alterity that come to disturb the self-certain subject. If we follow Derrida's lead, however, we should try to understand that this critique was above all meant to be a critique of the age-old presupposition that the 'animal other' is deprived of the *logos*.

[14] *Ibid.*, 27.

One of the first times Derrida explicitly addressed this issue was in an interview with Elisabeth Roudinesco. In this interview, Derrida argues that the relations between humans and animals must change drastically and adds that this is not merely an "ontological necessity," but also, and especially today, an "ethical duty."[15] In contrast to his brief and rather purely theoretical reflections on this issue in earlier texts, on this occasion and in *The Animal That Therefore I Am* he even goes as far as stating that "a war has been declared on so many animals,"[16] and that the violence inflicted upon them could even be compared "to the worst cases of genocide."[17] Many consider this comparison between the slaughtering of animals and the systematic liquidation of human beings totally unacceptable because it illegitimately equates the worst cases of human suffering and the treatment of animals in big industry. We would, however, completely misunderstand Derrida's intention if we take it to mean that he is blind to the fundamental differences between these two kinds of suffering, or even worse, that he abuses the suffering of those who were murdered in genocides to call attention to the suffering of animals. His point is not that we put too much weight on human suffering in comparison to animal suffering, but that we do not pay enough attention to the *singularity* of animal suffering. As Matthew Callarco has argued,

> The very difficult task for thought here is to bear the burden of thinking through both kinds of suffering in their respective singularity *and* to notice relevant similarities and parallel logics at work where they exist. To do so requires abandoning [...] the hierarchical humanist metaphysics that we have inherited from the ontotheological tradition [...].[18]

The common view that human life is intrinsically more valuable than animal life is deeply embedded in Western tradition and culture and it will therefore take more than rational argumentation and political commitment to bring about a radical reorientation of our value system. So although Derrida declares to have much sympathy for initiatives such as those of the animal rights movement,[19] he is far from convinced that the violence against animals could be stopped or lessened by simply granting them legally sanctioned rights. For not only is it exactly the Cartesian model of subjectivity upon which the modern concept of rights is founded

15 Jacques Derrida and Elisabeth Roudinesco, *For what tomorrow... A dialogue*, trans. Jeff Fort (Stanford: Stanford University Press, 2004), 64.

16 *Ibid.*, 67.

17 Derrida, *The Animal*, 26.

18 Matthew Callarco, *Zoographies. The question of the animal from Heidegger to Derrida* (New York: Columbia University Press, 2008), 112.

19 Derrida and Roudinesco, *For What Tomorrow*, 74.

that he considers to be responsible for justifying anthropocentric violence in the first place, but he also clearly states that he has "never believed in some homogenous continuity between what calls *itself* man and what *he* calls the animal."[20] This latter statement is clearly directed at animal rights theorists such as Peter Singer who argue that one should grant 'human' rights to those animals that are endowed with sufficiently developed cognitive capacities. Since Derrida takes this to imply that certain species of animals are to be considered superior to, for example, mentally-ill persons, he calls such strategies that deny any qualitative difference between humans and animals nothing less than "ridiculous and heinous."[21]

It should be clear, then, that Derrida takes issue with both dominant positions in the discussion of the human/animal relation. It is no surprise though that the largest part of his writings on this problem deal with the anthropocentric bias of the continental philosophical tradition. His main critique of this tradition is that it has always given humans a privileged position among the living creatures by endowing them with qualities, essences, or 'propers' such as language, consciousness, reason, or moral awareness which animals apparently do not possess. At the same time, however, he seems to be equally critical of those who strive to undo the metaphysical separation of humans and animals by giving back to the animal the qualities which the human has always denied it. According to Derrida, such a view does not take sufficiently into account the singularity of all living beings. In other words, Derrida aims to undo the radical human/animal binary, but at the same time he also wants to avoid putting man on the same ontological level as the animal. The question is, of course, whether it is actually possible to steer clear from both of these positions.

It seems as if the full import of this problem only started to dawn on him in the wake of his critical confrontation with Heidegger's thought. It is well-known that Heidegger was an ardent critic of the many forms of "ontotheological humanism" that populate the philosophical universe and it would therefore have seemed reasonable for Derrida to assume that Heidegger's writings contain a thinking that does not take for granted the radical human/animal divide. It is probably with this expectation in mind that in *Of Spirit* (1989) Derrida ventured a critical reading of Heidegger's only text that explicitly broaches the problem of the being of the animal, namely *The Fundamental Concepts of Metaphysics* (1995). As mentioned in chapter three, the guiding threat of Heidegger's argument in this lecture course is constituted by three theses: the stone is wordless, the

20 Derrida, *The Animal*, 30.

21 Derrida and Roudinesco, *For What Tomorrow*, 67.

animal is poor in world and man is world-forming. Derrida calls particular attention to the obscurity of the second thesis: what does being poor in world mean? Heidegger is quick to argue that this thesis certainly does not imply that there is a difference in degree between the human and the animal in the sense that the animal has only limited access to the entities in the world to which the human would have full access. Given that Heidegger clearly states that "every animal and every species of animal as such is just as perfect and complete as any other,"[22] it is certainly not the case that these characterizations should be taken as an evaluative ranking of some sort. Yet, the animal must certainly 'have' some world, otherwise it could not be distinguished from the stone of which he said that it has no world at all. The animal, Heidegger will conclude, has access to entities, but it does not have access to entities *as such*. Derrida comments as follows:

> This analysis, certainly, has the interest of breaking with difference of degree. It respects a difference of structure while avoiding anthropocentrism. But it remains bound to reintroduce the measure of man by the very route it claimed to be withdrawing that measure – this meaning of lack and privation.[23]

A lizard lying on a rock, to use one of Heidegger's famous examples, certainly relates to that rock in some way, but it certainly does not relate to the rock as such. The animal, Heidegger argues, is completely captivated [*benommen*] or absorbed by the entities which constitute its environment and therefore cannot have a free relationship to them. It always has a relation of utility with the entities it encounters and therefore cannot let the rock be what it is. Since the animal is captivated by its environment the possibility to apprehend something as something is structurally withhold from the animal.[24] This phenomenological possibility to reveal an entity in its being is reserved for Dasein only. Derrida concludes:

> I do not mean to criticize this humanist teleology. It is no doubt more urgent to recall that, in spite of all the denegations or all the avoidances one would wish, it has remained *up till now* [...] the price to be paid in the ethico-political denunciation of biologism, racism, naturalism, etc.[25]

The question is, then, whether Derrida himself found a way to avoid the twin pitfall of metaphysical humanism and biological reductionism. At the time of the writing of *Of Spirit* Derrida still seemed to be very

---

22 Heidegger, *Fundamental Concepts of Metaphysics*, 194.

23 Jacques Derrida, *Of spirit. Heidegger and the question*, trans. Geoffrey Bennington & Rachel Bowlby (Chicago: The University of Chicago Press, 1989), 49.

24 Heidegger, *Fundamental Concepts of Metaphysics*, 247.

25 Derrida, *Of Spirit*, 56.

skeptical about the prospects to actually achieve this goal. Heidegger's case is exemplary in this respect. Although he consistently rejected the reduction of Dasein to a biological entity as well as the metaphysical determination of the human as a rational animal, Derrida points out that in his notorious rectoral speech, 'The Self-assertion of the German university' (1993), Heidegger capitalizes on both evils at the same time by spiritualizing the forces of "earth-and-blood." He consequently ends up combining "the sanctioning of Nazism and the gesture that is still metaphysical."[26] Having become aware that it is simply impossible to twist oneself completely free from both humanism and biologism, the question for Derrida then becomes that of knowing which is the least grave form of complicity with these two 'evils.' In *Of Spirit* he does not yet elaborate such a negotiation with the anthropological limit, but he does give an indication of how one would probably have to proceed:

> [...] [C]an one not say just as legitimately that the having-a-world also has for man the signification of some *unheimliche* privation of world, and that these two values are not opposed?[27]

This is indeed the strategy that Derrida will pursue in *Aporias* (1993). In the second essay of this little book, 'Awaiting (at) the Arrival,' Derrida remind us that Heidegger opened *Being and Time* by arguing that human Dasein in the only entity in the world that has the ability to question its own being and that an analysis of its existentials provides a privileged starting point to reopen the question of being as such. In the second part of his *magnum opus*, Heidegger shows that Dasein's exceptionality resides in the fact that it is the only entity that has an experience of death. Only human Dasein can die, animals simply perish. Death, in other words, is Dasein's most proper possibility. But, Derrida objects, isn't death an experience that will never present itself to Dasein and thus actually its most improper possibility? No one will be able to experience his own death, no one will be able to 'live' this experience. From this perspective, the relationship between *Dasein* and the animal suddenly appears in a very different light:

> But if the impossibility of the "as such" is indeed the impossibility of the "as such," it is also what cannot appear as such. Indeed, this relation to the disappearing as such of the "as such" – the "as such" that Heidegger makes the distinctive mark and the specific ability of *Dasein* – is also the characteristic common *both* to the inauthentic *and* to the authentic forms of the existence

26 *Ibid.*, 40.

27 *Ibid.*, 50.

> of Dasein, common to all experiences of death and also, outside of Dasein, common to all living beings in general.[28]

Heidegger's metaphysical humanism is here defeated on its own terrain; Not by granting animals access to death 'as such,' but by showing that human *Dasein* does not have access to the 'as such' of death either.

Human finitude, Derrida recapitulates in *The Animal That Therefore I Am*, does not reside in man's ability to run ahead towards death, but in its not-being-able to experience death, in a lack of power, in a not-having of the 'as such.' This brings the human in close proximity to the animal. The most important question, Derrida argues, is therefore not whether animals do or do not have access to *logos*, speech, spirit, *tekhnē*, death, and the like, but to know whether animals can suffer. Being able to suffer is precisely this possibility without power, this radical finitude that the human shares with the animal.

At the time of the writing of *The Animal That Therefore I Am*, however, Derrida seems to have become aware of the fact that this line of thinking could also clear the way for a more subtle version of anthropocentric thought. This strand of anthropocentrism would not oppose the lack of the animal to the plenitude of the human, but claims that man's superiority over the animal derives precisely from his deficient nature or his originary default:

> [We will] see appear in every discourse concerning the animal, and notably in Western philosophical discourse, the same dominant, the same recurrence of a schema that is in truth invariable [...]: what is proper to man, his subjugating superiority over the animal [...] would derive from his originary fault, indeed from this default in propriety.[29]

The qualities that make the human superior to the animal, such as *logos*, mind, language, etc., are not simply gratuitous gifts from God or Nature, but emerge from the human's exceptional position as a deprived being. This would be the fundamental logic underpinning all metaphysical discourses concerning the human/animal divide. Unlike the animal, the human does not have a fixed nature and that would paradoxically explain why he is able to gain access to these various defining traits. Leonard Lawlor has argued that this schema can be found in nearly all traditions of Western thought, but that it is most clearly present in those traditions that assume the fallen nature of the human such as Platonism and Christianity:

---

28 Derrida, *Aporias*, 75.

29 Derrida, *The Animal*, 45.

> Because of a fault, man conceives animals as being absolutely innocent, prior to good and evil, "without fault or defect." The animals therefore do not seem to suffer a fall [...]. In contrast, man is not perfect; he has fallen and has a fault, which allows him to question. The ability to question brings us to the axiom: it is precisely "a fault or defect" in man, in us, that allows us to be masters over the animals.[30]

The human would have suffered a fatal fall as a result of which he paradoxically attains a privileged position among the living beings, "[t]his would be the law of an imperturbable logic, both Promethean and Adamic, both Greek and Abrahamic (Judaic, Christian, Islamic)."[31] Whence this sudden change of discourse? It almost seems as if Derrida's argument that the human also suffers from a deficient constitution, that man is also "poor in world," had somehow started to boomerang back on him so that he deemed it necessary to ward off the suspicion that his own writings would be open to such an interpretation.

Epiphylogenetic Life

There is little doubt Derrida was mainly targeting Stiegler's theory of technics with his critique of those who hold that what is proper to the human derives from an "originary fault." Since Stiegler wrote his first volume of *Technics and Time* as a doctoral student of Derrida, the two were very familiar with each other's work and it would therefore certainly not be impossible that they disagreed on certain issues. There is not only biographical evidence that there was a dispute going on between these two deconstructive thinkers though. In the already mentioned essay 'Quand faire c'est dire,' Stiegler explicitly explained that his theory should be understood as an attempt to resolve the question of the human/animal distinction which he thought Derrida had not adequately addressed. In this article, Stiegler returns to Derrida's reading of Heidegger in *Of Spirit* which we discussed earlier. If the human is also "poor in world" and the human's relation to the world is equally "guided by a narrow sphere of drives, of desires"[32] and the like, Stiegler argues, then there still remains to be determined what constitutes the human mode of being poor in world. In other words, if the human is also a programmed being, if he does not have unmediated access to the world either, then the specificity of the human program still remains to be thought. Indeed, Derrida's argument only seems to beg the question, for if the human and the animal

30 Leonard Lawlor, *This is Not Sufficient. An Essay on Animality and Human Nature in Derrida* (New York: Columbia University Press, 2007), 67.

31 Derrida, *Animal*, 20.

32 *Ibid.*, 159.

share a radical powerlessness, does he not end up endorsing some kind of continuism after all? Derrida seems to have anticipated such objections and counters them as follows:

> [...] I am not advocating the blurring of differences. On the contrary, I am trying to explain how drawing an oppositional limit itself blurs the differences, the *différance* and the differences, not only between man and animal, but among animal societies – there are an infinite number of animal societies, and, within the animal societies and within human society itself, so many differences.[33]

But even if it is indeed so that the philosophical tradition has always treated 'the animal' as one homogenous category and so neglected the singularity of animal life, then we still bear the burden to determine what constitutes the singularity of the human animal. In order to steer clear from the traditional metaphysical prejudices, Stiegler will answer this question by showing that the transition from the animal to the human is not the transition from a fully programmed living being to a being guided by no program whatsoever, but the transition from a genetic program to a techno-logical or what he calls *epiphylogenetic* program.

In the beginning of the third chapter of *The Fault of Epimetheus* Stiegler writes that what is at stake is this book is "to think the birth of the human [...] and to attempt this independently of all anthropologism."[34] Stiegler finds a suitable point of departure for this project in the work of the French paleontologist André Leroi-Gourhan. What is particularly interesting for Stiegler is that Leroi-Gourhan describes anthropogenesis as a process of "exteriorization" in the course of which the brain and the tool develop at equal pace. As this non-anthropocentric concept defies the classical image of the human as an animal to which miraculously the capacity for reflection is added, it seriously calls into question the radical opposition between the human and the animal. A crucial event for the development of Leroi-Gourhan's theory was the discovery of the remains of the Australopithecus together with its flint tools in 1959. According to Leroi-Gourhan, this was the first bipedal humanoid to have benefited from exteriorization. What is so important about this paleontological finding is that the small size of the Australopithecus' brain pan urges the conclusion that what was decisive in anthropogenesis was not cerebral development, but the erect posture and the new functional organization of the body that ensued from this novelty in the history of life:

---

33 Jacques Derrida (1987), 'On reading Heidegger: an outline of remarks to the Essex colloquium,' *Research in Phenomenology* 17: 183.

34 Stiegler, *The Fault of Epimetheus*, 135.

> The "freeing" of the hand during locomotion is also that of the face from its grasping functions. The hand will necessarily call for tools, movable organs; the tools of the hand will necessarily call for the language of the face. The brain obviously plays a role, but it is no longer directive: it is but a partial element of a total apparatus, even if the evolution of the apparatus tends towards the deployment of the cerebral cortex.[35]

The erect posture brought about a gradual suspension of genetic constraints and, consequently, an increasing indetermination of body functions. The immediate result was the appearance of tools and language to fill in for this lack of genetic programmability. Moreover, Stiegler writes, between the Australopithecus and the Neanderthalian tools evolved so slow "that one hardly can imagine the human as its operator, rather, one much more readily imagine the human as what is invented."[36] This will be Stiegler's thesis: the birth of the human *is* the appearance of technics in the sense that the human and technics "invent each other respectively, as if there were a technological maieutic of what is called 'humanity,'"[37] the cortex being informed by the tool as much as the tool by the cortex.

Leroi-Gourhan, however, will maintain that from the Australopithecus up to the Neanderthalian, "tools were still, *to a large extent*, a direct emanation of *species* behavior,"[38] which means that he thinks that in this archaic period of thousands of years technical evolution was still mainly determined by the rhythm of cortical development, itself propelled by genetic selection, and thus essentially of zoological origin. According to the paleontologist it is only with the emergence of the Neanderthalian that technical evolution frees itself completely from genetic constraints. So although Leroi-Gourhan argues that the fact that the Australopithecus could already handle tools implies that this humanoid must have had a "technical consciousness," he thinks he was still not yet endowed with a full-fledged "symbolic consciousness," which will only emerge with the Neanderthalian when an extraordinary increase in anticipatory capacities was accompanied with the stabilization of the evolution of the human's neuro-equipment. What is at issue here is the latitude that became available with exteriorization and the origins thereof. Stiegler argues that Leroi-Gourhan's distinction between two types of 'consciousness' contradicts the simple fact that the Australopithecus must already have had the capacity for anticipation because the possibility to perform a technical

35 *Ibid.*, 145.

36 *Ibid.*, 134.

37 *Ibid.*, 142.

38 André Leroi-Gourhan, *Gesture and speech*, trans. Anna B. Berger (Cambridge Mass.: MIT Press, 1993), 97.

gesture requires, as Leroi-Gourhan himself explains, "a good deal of foresight on the part of the individual performing the sequence of technical operations."[39] So although Leroi-Gourhan came close to recognizing that the birth of the human coincides with the birth of technics, in the end he gives a determining role to cortical development, implying that the Australopithecus was still an animal and that the real origin of the human is the acquisition of a faculty for symbolization. This is a conclusion which Stiegler cannot accept:

> There is no such [second] origin because technical differentiation presupposes full-fledged anticipation, at once operative and dynamic, from the Australanthropian onwards, and such anticipation can only be a relation to death, which means that symbolic intellectuality must equally be already there. Reflective intellectuality is not added to technical intelligence. It was already its ground.[40]

The difficulty Leroi-Gourhan found himself confronted with, Stiegler argues, is the ambiguity of the term "exterior-ization." It gives the impression that what is exteriorized should be preceded by some kind of (mental) interiority, something that would function as the origin of this movement. But given the fact that Leroi-Gourhan himself had asserted that the brain was only a beneficiary of the rupture of exteriorization, there cannot be anything of that kind, whether it be in the guise of a neurological impetus or a premature consciousness. What has therefore remained unthought-of is the possibility that the "interior and exterior are [...] constituted in a movement that invents both one and the other,"[41] so that "neither one precedes the other, neither is the origin of the other, the origin being the coming into adequacy [*con-venance*] or the simultaneous arrival of the two."[42] The challenge is to think what Stiegler calls a "mirror proto-stage" or an "instrumental maieutics," the structural coupling of the human and technics that makes the constitution of the one impossible and unthinkable without the other.

It is at this stage of his argument that Stiegler calls upon Derridean *différance* which in this context he understands as the history of life in which are articulated different stages of the *grammē*. Significantly, Derrida also based his concept of the *grammē* on what Leroi-Gourhan calls the liberation or exteriorization of programs. More specifically, Derrida finds in Leroi-Gourhan's notion of "program" a "non-anthropocentric concept

---

39 *Ibid.*, 97.

40 Stiegler, *The Fault of Epimetheus*, 163.

41 *Ibid.*, 142.

42 *Ibid.*, 152.

that does not take for granted the usual divides between animality and humanity."[43] Since Derrida's interpretation of Leroi-Gourhan was crucial for Stiegler, I quote him at length:

> Leroi-Gourhan no longer describes the unity of man and the human adventure thus by the simple possibility of the graphie in general; rather as a stage or an articulation in the history of life – of what I called *différance* – as the history of the *grammē*. [...] This movement goes far beyond the possibilities of the 'intentional consciousness.' It is an emergence that makes the *grammē* appear as such. [...]. If the expression ventured by Leroi-Gourhan is accepted, one could speak of a 'liberation of memory,' of an exteriorization always already begun but always larger than the trace which, beginning from the elementary programs of so-called 'instinctive' behavior up to the constitution of electronic card-indexes and reading machines enlarges *différance* and the possibility of putting in reserve.[44]

Since the *grammē* does not only concern genetic inscription, but also "electronic card indexes" and "reading machines," that is, means of technical inscription, it seems that the program is operative on both sides of the human/animal divide. But, Stiegler argues, it is nonetheless true that this transition from a genetic to a non-genetic program coincides with the appearance of the human and an account of the specificity of this passage is still missing in Derrida's writings.

The emergence of the human, Stiegler will conclude, coincides with the appearance of a third type of memory or program in the evolution of the living. If animal life already knows genetic memory and epigenetic memory, with the birth of the human, there emerged a third kind of memory which Stiegler calls epiphylogenetic memory – memory embodied in technical objects. Since the human is capable of exteriorizing its epigenetic memory onto material supports – from flint tools to digital storing devices – it can preserve and transmit this memory across generations. It is this suspension of the law of life that Stiegler situates the singularity of the human animal.

Critics such as Ben Roberts (2005) and Geoffrey Bennington (1996) have argued, however, that it is not so much Derrida's interpretation of Leroi-Gourhan that is lacking in clarity, but Stiegler's appropriation of it. While Derrida's quasi-concepts such as the *grammē* and the trace precisely challenge the opposition between the human and the animal, Stiegler's distinction between epigenesis and epiphylogenesis would simply reproduce this opposition in terms of an originary technicity. By insisting on

43 *Ibid.*, 137.

44 Jacques Derrida, *Of grammatology*, trans. Gayatri C. Spivak (Baltimore: John Hopkins University Press, 1974), 84.

the fact that the human is the only living being that is able to conserve its memory in technical objects, Stiegler would merely have invented a more subtle form of anthropocentrism. As we have seen, this also seems to be Derrida's own critique of Stiegler's theory of anthropogenesis.

We would, however, do injustice to Stiegler's theory if we read it as simply implying that technics is the defining property of man since his argument is precisely that there is nothing proper to the human. If the danger of anthropocentrism is most critical when it straightforwardly opposes the animal to the human, then no worse could be done than to install a insurmountable gap between them on the basis of some transcendental property. In this respect Stiegler's refutation of what he calls the "miracle thesis," that is, the positing of some form of 'spirituality' without a rational understanding of its provenance, is rather an attempt to diminish anthropocentric violence than its continuation by other means. Moreover, the suppression of both overt and latent anthropocentrism could just as much lead to its opposite, namely the effacing of all differences between the animal and the human. As Stiegler puts it, "the contestation of oppositions must not eliminate the genesis of differences."[45]

> When asked in an interview what he thinks about the fact that certain primates also use tools and transmit this knowledge to the next generation, Stiegler responds as follows:
>
> If you would object to me that certain large apes also have cultures, then I would say that I am willing to accept them as members of the world which starts with the human – in other words, as embryonic fabricators of this third type of memory. I would most certainly allow them to enter human history. As a matter of fact, that is the reason why they are so close to us.[46]

Arthur Bradley has argued that this way of responding to the above objection can only confirm the impression that Stiegler seeks to violently "absorbs every apparent exception into the narrative of hominization."[47] We would argue, however, that such an interpretation can only arise because it is almost impossible to use the name of the "human" without inheriting the metaphysical assumptions pertaining to this term. That is the reason why Heidegger uses the term Dasein to designate human being and that is also why Stiegler prefers to call this entity simply the "who." In other words, Stiegler is not primarily interested in the entity we unreflectively call the 'human,' but in the structure of experience that opens

---

[45] Stiegler, *The Fault of Epimetheus*, 163.

[46] Stiegler, *Philosopher par Accident*, 49; Translation mine.

[47] Arthur Bradley, 'Originary technicity? Technology & Anthropology,' in *Technicity*, ed. A. Bradley and L. Armand (Prague: Litteraria Pragensia, 2006), 98.

up when a living being enters into a mutual constitutive relation with technical objects:

> I'm not interested in what we call "man" but in *temporality*, temporality that, as a form of life, has to decide what it is to become (and it so happens that this form of life is still called man today). Even when man is finished, when he belongs to the past, this form of life may well continue on, becoming ever more complex—and perhaps man is *already* finished.[48]

Thinking anthropogenesis as mutually constitutive with technogenesis implies that the human is not a spiritual miracle that is added to an animal body, but that hominization is "the pursuit of the evolution of the living by other means than life."[49] Hominization is not a break *with* life, but a break *in* life and, moreover, *through* technology.

Technology and Communicative Reason

As its title clearly suggests, Stiegler's project to forge a new conception of the human/technology relationship results not only from a critical dialogue with Derrida on the question of the anthropological difference, but also with the work of Heidegger on the constitution of human temporality. Stiegler's attempt to think technics as constitutive of temporality will lead him to articulate a fierce critique of the latter's existential thinking of *Technik*. Since Heidegger opposes the primordial temporality disclosed in *Sorge* [care] to the technical constitution of temporality disclosed in *besorgen* [concern], Stiegler's believes Heidegger's work is still caught in the classic metaphysical tradition that sees technology as external to human nature. In *Being and Time*, the temporality of concern marks the mode of existence of *das Man* [the One] whose inauthentic understanding of Being ensues from the uprooting consequences brought about by modern industrial society. Stiegler, on the other hand, argues that technics is precisely the very condition of the experience of time since technical objects provide the conditions for Dasein to experience a past that one has not lived and to anticipate the future in the light of this inherited past. According to Stiegler, Heidegger was very close to grasping this fundamental relationship between technics and time when he argued that worldhood is primarily disclosed in the handling of equipment [*das Zeug*] that is ready-to-hand [*Zuhandensein*], but, as Richard Beardsworth has argued recapitulating Stiegler's critique, Heidegger refrained from

48 Bernard Stiegler (2003), 'Technics of decision,' *Angelaki* 8 (2), 158.

49 Stiegler, *The Fault of Epimetheus*, 135.

drawing this radical conclusion in his desire to "understand the factuality of the world in terms of Dasein's self-affection."[50]

Yet, notwithstanding the predominance of references to Heidegger's work throughout his writings, there are strong indications that in *The Fault of Epimetheus* Stiegler also envisages a critical assessment of Habermas's theory of technology. However, a comprehensive elaboration of such a critique is noticeably missing. Habermas's seminal essay 'Technology and Science as "Ideology"' is briefly mentioned and criticized by Stiegler in the introduction of the book, but his attention is soon drawn to Heidegger's work.[51] This is all the more curious, since Stiegler explicitly states that "the major themes from the existential analytic will be interpreted and submitted to a critique that I sketched out as much in response to Habermas as to Heidegger."[52] Although Stiegler does indeed indicate that Habermas and Heidegger share a similar pessimistic diagnosis of modern technology by saying that "Habermas and Heidegger appear to agree in considering the technicization of language as a perversion,"[53] it is clear that his critique of Heidegger cannot be transposed to Habermas's work: arguably, existential ontology and critical theory are not kindred spirits. We will, therefore, aim at developing a Stieglerian reading of Habermas's writings on technology and argue that the opposition the latter draws between technics and language, and which appears under such different terms as work/interaction and purposive-rational action/communicative action, stems from his failure to fully confront what Stiegler calls the "techno-logical condition of man."[54]

Habermas's 'Technology and Science as "Ideology"' was written in direct response to Herbert Marcuse's *One-Dimensional Man* (1964). In earlier works, like *Eros and Civilization* (1955), Marcuse had tempered the pessimism of his Frankfurter School colleagues about a society dominated by instrumental reason by pointing out that the increasing automation of the labor-process could also create a realm of freedom in which life-enhancing activities would thrive. In *One-Dimensional Man*, however, this initial optimism concerning technological development makes way for radical critique and blatant utopian hope. Marucse frames technology and science as the media through which the dominant political ideology exercises and strengthens its power. Marcuse no longer believes that the technology can be seen as a neutral means, as classic Marxism

50 Beardsworth, *From a genealogy of matter*, 152.

51 See Stiegler, *The Fault of Epimetheus*, 10-13.

52 *Ibid.*, 16-17.

53 *Ibid.*, 13.

54 *Ibid.*, 15.

had always maintained, because its drive to dominate nature and humanity is reproduced on the political level where its rationality serves to justify stringent forms of social control. That is, although the technological system is structurally repressive, humanity does not experience this as such because the system is legitimized by the comforts of life it yields. Therefore, in Marcuse's view, the modern technological system can no longer fulfill its role as the material basis of a liberated society, but needs itself to be liberated for such a society to be possible in the first place. Hence Marcuse's curious call for a "New Technology" which would treat nature no longer as inert matter to be dominated at will, but as a subject in its own right, and a corresponding "New Science," which "would arrive at essentially different concepts of nature and establish essentially different facts."[55]

In 'Technology and Science as "Ideology",' Habermas takes up Marcuse's thesis of the ideologization of science and technics, but he suggests that a distinction must be made between the "institutional framework" and the "subsystems of purposive-rational action." Both structures are subject to waves of rationalization, but while the institutional framework finds its source of legitimation in communicative processes – from mythological explanations to human rights claims –, progress in the subsystems of purposive-rational action is measured by the degree in which its practices prove to be successful or not. On the basis of this distinction, Habermas seeks to explain a phenomenon that Marcuse and the other members of the *Frankfurter Schüle* described either as the dialectic of enlightenment, the eclipse of reason or one-dimensionality, while denying that pessimism or a utopian call for a "New Technology" are the only responses left. Habermas's strategy consists of showing that there is nothing wrong with the increasing rationalization of the subsystems of purposive-rational action as such, but that in modern society its logic has also penetrated the institutional framework. The result is that issues that ought to be decided upon by citizens following communicatively guided norms are being increasingly treated as mere technical ones. Hence, the problem for Habermas is not modern science and technology *per se*, but the fact that "the reified models of the sciences migrate into the sociocultural life-world and gain objective power over the latter's self-understanding."[56] Therefore, the solution must rather be sought in "removing the restrictions on communication."[57] It is in this sense that we ought to understand

55 Herbert Marcuse, *One-Dimensional Man* (Boston: Beacon Press, 1964), 166-177.

56 Habermas, *Technology and science as "ideology"*, 113.

57 *Ibid.*, 118.

Habermas's claim that the aim of Marcuse's "New Technology" – the liberation of nature – is utopian:

> [...] [T]he achievements of technology, which are indispensable as such, could surely not be substituted for by an awakened nature. The alternative to existing technology, the project of nature as an opposing partner instead of object, refers to an alternative structure of action: to symbolic interaction in distinction to purposive-rational action.[58]

Put differently, Habermas criticizes the modern tendency to apply the logic of instrumental-technical action to the symbolic driven sphere of culture, as well as Marcuse's suggestion to do the exact opposite, namely to apply concepts that only make sense in the realm of intersubjectivity (domination, liberation, opposing partner, etc.) to the realm of nature where they make no sense at all.

After a brief exposition of the discussion between Marcuse and Habermas, Stiegler remarks that since Habermas considers the technicization of the sphere of communication as a denaturalization, "as it were a question of one instance 'proper to humanity' perverting another instance 'proper to humanity',"[59] his analyses are still haunted by the founding positions of metaphysics. According to Stiegler, the technicization of language was already at issue in the conflict between Plato and the sophists. In the *Phaedrus* and many other dialogues, Plato criticizes the sophists' use of the technique of writing as a memory aid (*hypomnēsis*). According to Plato, the sophists were only able to become excellent public speakers because they wrote down their speeches in advance, which allowed them to learn everything by heart. Plato thinks this a dangerous practice because, in his view, writing is a *pharmakos*, a poison that ruins the truth seeking capacity of living memory (*anamnēsis*). The technique of writing distorts truth because it replaces the rule of the *logos* by that of death letters and in this way renders a critical discussion of the propositions under discussion well-nigh impossible. For Plato, true *epistēme* is only possible through a spoken dialogue that fixes attention on the ideal content of propositions. In this sense, Stiegler argues, we encounter here at the heart of critical theory a reintroduction of Plato's lament that the technicization of language does harm to individuation:

> The new ideology consequently violates an interest grounded in one of the two fundamental conditions of our cultural existence: in language, or more

---

58 *Ibid*., 88.

59 Stiegler, *The Fault of Epimetheus*, 13.

precisely, in the form of socialization and individuation determined by communication in ordinary language.[60]

According to Stiegler, it is no coincidence that this age-old discussion should rise up again today. Habermas's essay was written in a period in which one was still trying to come to terms with a new crisis concerning the structure of memory that appeared in the wake of the Industrial Revolution, and thus when the technical system "suddenly outgrows the sphere of language – which is also to say the *logos* – and takes over that of material bodies."[61] According to Stiegler, the Industrial Revolution can be described as a new stage in a process of exteriorization in which the memory of living bodies is inscribed in machines, in the sense that the "the memory of the proletarian has been absorbed by a machine reproducing gestures that he no longer needs to know how to make, and which he must now simply serve, because he has reverted to the status of a serf."[62] In the industrial stage of exteriorization, new powers are bestowed upon the human, but also new threats. Habermas implicitly mentions this process in his attempt to counter Marcuse's argument for a New Technology. While Marcuse argues that the contemporary technical system is historically contingent, Habermas argues that this is not the case, given that technological development is a 'project' of the human species as a whole:

> In any case technological development lends itself to being interpreted as though the human species had taken the elementary components of the behavioral system of purposive-rational action, which is primarily rooted in the human organism, and projected them one after another onto the plane of technical instruments, thereby unburdening itself of the corresponding functions. At first the functions of the motor apparatus (hands and legs) were augmented and replaced, followed by energy production (of the human body), the functions of the sensory apparatus (eyes, ears, and skin), and finally by the functions of the governing center (the brain).[63]

Habermas describes technical evolution as a process of exteriorization, but not in Stiegler's sense of a mutually constitutive relationship between the human and technics whereby the one requires the other to be itself in the first place, but as a structural 'emptying-out' of the human behavior system, the functions of which are progressively projected

---

60 Habermas, *Technology and science as "ideology"*, 113.

61 Bernard Stiegler, 'Anamnēsis and Hypomnēsis: The Memories of Desire,' in *Technicity*, eds. A. Bradley & L. Armand (Prague: Litteraria Pragensia, 2006), 20.

62 *Ibid.*, 21.

63 Habermas, *Technology and science as "ideology"*, 87. See also Jürgen Habermas, *Knowledge and Human Interests*, trans. J. J. Shapiro (London: Heinemann, 1972), 48.

into technical objects. Hence, Habermas defines technical objects as mere material extensions of biological functions that increase the human's power over its external environment. It is this instrumental conception of technics that motivates Habermas to criticize the technicization of the institutional framework in modern society: instead of fulfilling its proper function as a means at the disposal of man, man himself has now reverted to the status of raw material for the technical system:

> For the first time man can not only, as *homo faber*, completely objectify himself and confront the achievements that have taken on independent life in his products; he can in addition, as *homo fabricatus*, be integrated into his technical apparatus if the structure of purposive-rational action can be successfully reproduced on the level of social systems.[64]

Given that, in Habermas's view, "individuation is achieved through the socializing medium of thick linguistic communication,"[65] the technicization of society corrupts this essential human property. The crucial question for Stiegler, however, is "whether such an evaluative distribution – according to which technics remains only *on one side* (of an opposition), itself not *constitutive* of individuation – in fact remains 'metaphysical'."[66] Insofar Stiegler thinks that technics constitutes the condition of possibility for the capacity for memorization and anticipation, the becoming of the "who," is as much conditioned by technics as it is by language.

In order to get to the root of this issue, we have to turn to Habermas's book *Knowledge and Human Interests* (1972) in which the theoretical basis for his critique of Marcuse was formulated. In this early work, Habermas distinguishes between two fundamental modes of human action, 'work' and 'interaction.' This distinction corresponds to the division between the subsystems of purposive-rational action and the institutional framework.[67] These modes of action are, in turn, linked to a particular human interest that serves as the basis of a specific form of knowledge. On the one hand, work – or instrumental action – is linked to the technical interest in the mastery of nature. This interest is necessary for the reproduction of the species and, moreover, lies at the basis of the empirical-analytical knowledge of the natural sciences. Interaction – or symbolic action –, on the other hand, is linked to the fundamental interest

64 Habermas, *Technology and science as "ideology"*, 106.

65 Habermas, *The Future of Human Nature*, 54.

66 Stiegler, *The Fault of Epimetheus*, 14.

67 Habermas also recognizes an emancipatory interest in knowledge. But since this third interest is not grounded in deep-rooted anthropological structures, he thinks it has a derivative status. See for example Jürgen Habermas, 'A postscript to *Knowledge and Human Interests*,' *Philosophy of Social Science* 3, 176.

in maintaining distortion-free communication. This interest is necessary for the reproduction of cultural existence and lies at the basis of the historical-hermeneutic knowledge of the cultural sciences. On the basis of this theory of "equiprimordial" knowledge-constitutive interests, Habermas can retain important elements from Marcuse's critique of instrumental reason, while criticizing its most problematic claims. That is, while he agrees with Marcuse that instrumental action is not a neutral practice in that it serves a particular human *interest*, he rejects the possibility of a New Science and Technology by specifying that this interest is not bound to a particular social group but is shared by the human species as a whole. Since this fundamental mode of action is built into the very biological structure of the human species itself, the human can only perceive nature as a possible object of technical control. In other words,

> [t]hey have a transcendental function but arise from actual structures of human life: from structures of a species that reproduces its life both through learning processes of socially organized labor and processes of mutual understanding in interactions mediated in ordinary language. These basic conditions of life have an interest structure. The meaning of the validity of statements derivable within the *quasi-transcendental* systems of reference of processes of inquiry in the natural and cultural sciences is determined in relation to this structure.[68]

This passage contains a curious mixture of naturalistic and transcendental arguments. Before taking a closer look at how these relate to each other, we will first briefly examine them separately. Since Habermas wants to avoid the relativism of Marcuse's position, he first seeks to argue that the human interests are grounded in deep-seated anthropological structures. For this purpose, he finds an ally in Marx who, in opposition to Hegel, holds that "nature [...] not only seems external to a consciousness that finds itself within nature but refers instead to the immediacy of a substratum on which the mind contingently depends.[69] Whereas Hegel considers nature as a self-created presupposition of *Geist*, Marx argues that nature must be seen as an autonomous entity that gradually gave rise to both the human species and his external environment. Habermas similarly argues that the fundamental modes of human action and their corresponding knowledge-constitutive interests have emerged through basic evolutionary processes and assigns a pivotal role in to a rupture in bodily organization. Significantly, he suggests that the invention of the tool was an immediate consequence of this rupture:

68 Habermas, *Knowledge and Human Interests*, 194-195, Emphasis added.

69 *Ibid.*, 26.

> Without the particular physical equipment of the hominids, the 'process of material exchange' could never have assumed the form of labor at the human level. Men 'begin to distinguish themselves from animals as soon as they begin to produce their means of subsistence, a step that is conditioned by their bodily organization.' [...] The absolute ego of social production is founded in a history of nature that brings about the tool-making animal as its result.[70]

As in Leroi-Gourhan, hominization is described as a process inherently linked to the evolution of tools. This leads Habermas to the conclusion that it is impossible to determine the origin of man in either biological or transcendental terms:

> In contrast to the fleeting aspects of individual performances, productions, and gratifications, labor processes give rise to something general that accumulates in the productive forces. In their turn these enduring productions, or stored up forces of production, transform the world within which subjects relate to their objects. Therefore the species can have *no fixed essence*, either as a transcendental form of life or in the empirical form of a biologically conditioned basic pattern of culture.[71]

Technology, or the "productive forces," as Habermas calls it in a Marxist parlance, contains the accumulated materialized experiences of previous generations. What is transmitted through time is an ensemble of technical objects, which gives man access to an each time different, but always already meaningful world.

This seems to bring Habermas close to the idea that technology is a fundamental condition of human culture. However, there is yet another theoretical issue he wants to address and which leads him in a different direction. The reason why Habermas understands the activity of "work" also as a *transcendental* structure is that in *Knowledge and Human Interests* he is mainly concerned to counter the scientism of those positivists who hold the objectivist illusion that science investigates reality as such. By arguing that the reality that science takes as its object of inquiry is not an ontologically independent realm of facts, but that part of reality that is transcendentally constituted by our fundamental interest to master nature, Habermas is able to show that while science has an interest structure, and is therefore not value-neutral, it is nevertheless valid when applied within its own realm.

Ingenious as this theoretical construction may be, in the sense that it allows him both to dispel the positivistic account of science and discredit Marcuse's longing for a New Technology, the combination of both

70 *Ibid.*, 41.

71 *Ibid.*, 30-31, Emphasis added.

naturalistic and transcendental arguments leads to some serious contradictions. This is so because Habermas seeks to explain the *genesis* of human interests, while also retaining their *transcendental* character. Our knowledge of the phenomenal world is the necessary product of our transcendental interest in the mastery of nature. However, because Habermas also maintains that this interest is the product of contingent evolutionary processes, he has to assume the existence of a problematic pre-human *nature in itself* that, as it logically precedes the transcendentally constituted *nature for us*, functions as an absolute origin in his theory:

> While epistemologically we must presuppose nature as existing in itself, we ourselves have access to nature only within the historical dimension disclosed by labor processes. Here nature in human form mediates itself with objective nature, the ground and environment of the human world. "Nature in itself" is therefore an abstraction, which is a *requisite of our thought*.[72]

Clearly, there is a striking circularity at work here, for how is knowledge of a "nature in itself" possible given that he also holds that knowledge is always *constituted* through the fundamental interests?[73] As Thomas McCarthy puts it:

> In Habermas's terms, the categories and methods of empirical-analytical science would both be explained (transcendentally) by reference to structures of human action and be employed to explain (empirically?) these structures [...]. It seems to follow that nature is the ground of subjectivity. And this is, on the face of it, flatly incompatible with its status as a constituted objectivity.[74]

According to Habermas, there is clear evidence that our interest structures have a purely transcendental character: we can only perceive the world as an object of possible technical control. Yet, there is also a fundamental intuition that nature is something that predates us as a species and somehow retains its independence. By accepting both views at the same time, however, Habermas appears to be saying that the knowledge-constitutive interests are both constitutive of and constituted by nature. This problem has of course haunted philosophy ever since Kant's Copernican revolution, but whereas Kant keeps the phenomenal and the noumenal realms rigidly separated, Habermas presumes that he can solve this problem by combining Kant's theory of passive synthesis with Marx's view that human understanding also has an active element, which, in the *Theses on Feuerbach*, Marx

72 Habermas, *Knowledge and Human Interests*, 34. Emphasis added.

73 For a similar critique, see for example Joel Whitebook (1979), 'The Problem of Nature in Habermas,' *Telos* 40, 41-69 and Thomas McCarthy, *The Critical Theory of Jürgen Habermas* (Cambridge Mass.: MIT Press, 1981), 110-125.

74 McCarthy, *The Critical Theory of Jürgen Habermas*, 111-112.

calls "human sensuous activity." According to Habermas, nature in itself is positively knowable, but only as an object of possible technical control for a species that is compelled to reproduce its life through purposive-rational action. Therefore, the ultimate touchstone of this scheme is the obstinacy of material reality to which our behavior and our beliefs have to adapt:

> We do reckon with the existence of a reality that is independent of men who can act instrumentally and arrive at a consensus about statements. But what the predication of properties catches "of" this reality is a matter of fact that is *constituted only* in the perspective of possible technical control.[75]

Habermas argues here that nature in itself is an independent reality that somehow functions as the *unchanging* material ground of our experiences. However, the problem is, as we mentioned above, that he also wants to maintain that technology always "transform the world within which subjects relate to their objects"[76] First, does this initial view not imply that there is not one way of relating to the world – as a possible object of technical control – but that there are many such ways depending on the stage that technical evolution might have reached? This is also Heidegger's position. What Habermas considers as a transcendental way of relating to the world appears in Heidegger's later writings as something specific to modern society in which everything, including the humans themselves, is disclosed as a standing reserve [*Bestand*] that can be stockpiled and used to serve the will to power. Second, does his initial view not also imply that there is not one unchanging 'reality,' but that every time a new stage in technical evolution is reached, also a 'new' world is created? A hypothetical visitor from the 17th century would surely no longer recognize *our* world, with all its technical objects, as his world. This seems to be self-evident, but it is not for Habermas who, in the context of the discussion that arose in the wake of the publication of Thomas Kuhn's *The Structure of Scientific Revolutions*, maintains that "it is always the experience of identical objects of *our* world which is being interpreted differently according to the state of scientific progress we happen to have reached."[77]

Steven Vogel has suggested that this problem of a nature in itself in Habermas's theory of knowledge constitutive interests is the outcome of an act of "bad faith" on Habermas's part. Vogel argues that Habermas defined the human interests as species-related and not socially constructed because he wanted to avoid at any costs any "Lysenkoist" implications to

75 Habermas, *Knowledge and Human Interests*, 130.

76 Habermas, *Technology and science as "ideology"*, 29.

77 Habermas, *A postscript to Knowledge and Human Interests*, 171.

follow from his views.[78] This is an interesting hypothesis, but we want to suggest that it also follows from his disavowal of the techno-logical constitution of the human. By arguing that the technical interest of mastering nature is 'natural,' in the sense that it is vital for the reproduction of the species, Habermas may be able to avoid the danger of relativism that is present in Marcuse's writings, but the result is that he ends up maintaining that technics merely serves the instinct for survival. Habermas would probably deny that his theory includes the idea that instrumental action is propelled by a purely zoological drive because he explicitly argued that "the human interests that have emerged in man's natural history [...] derive both from nature and *from the cultural break* with nature,"[79] but this argument is not open to him because he almost immediately adds that "what raises us out of nature is the only thing whose nature we can know: *language*."[80] We meet once again with Leroi-Gourhan's concern to avoid the conclusion that what is specifically human is technicity. But whereas the paleontologist ends up installing an enigmatic gap between a technical pre-humanity and a symbolic humanity, in Habermas this gap reappears on the societal level due to his concern to safeguard the sphere of symbolic interaction from pervasive technology.

Inevitable Enhancement

It can be argued that Habermas's criticism of the colonization of the communicatively structured 'life-world' by systems of instrumental rationality is also at the basis of his rejection of any genetic intervention in the human. Because if we were to reduce Habermas's argument against genetic intervention to its most basic principle, then we would say that it holds that it is wrong because it threatens to destroy the whole meaning of what is to be human. That is to say, it assumes that the human has an immutable 'nature' and that any attempt to improve upon it through technology could be seen as a potential moral pitfall:

> Bodies stuffed with prostheses to boost performance, or the intelligence of angels available on hard drives, are fantastic images. They dissolve boundaries and break connections that in our everyday actions have up to now seemed to be of almost transcendental necessity. There is fusion or the organically grown with the technologically made, on the one hand, and separation of the productivity of the human mind from live subjectivity, on the other hand. Whether these speculations are manifestations of a feverish imagination or

78 Steven Vogel, *Against Nature: The Concept of Nature in Critical Theory* (Albany: SUNY Press, 1996), 122.

79 Habermas, *Knowledge and Human Interests*, 312.

80 *Ibid.*, 314.

> serious predictions, an expression of displaced eschatological needs or a new variety of science-fiction science, I refer to them only as examples of an instrumentalization of human nature initiating a change in the ethical self-understanding of the species – a self-understanding no longer consistent with the normative self-understanding of persons who live in the mode of self-determination and responsible action.[81]

Much of this argument depends on the distinction between the natural and the artificial, the "grown" and the "made," which Habermas sees as a "transcendental necessity." While inanimate nature is mute and passive and hence open to forms of instrumental manipulation, organic life-forms possess an inner dynamic of spontaneous growth that ought to put categorical restrictions on the kinds of actions we are allowed to perform on them. To be more precise, since human beings come into being by nature, they are entitled to "respect" or "dignity," and this should act as a check upon our practical dealings with them. However, as we have seen, Habermas denies that dignity is a property one possesses simply by virtue of being human, and instead claims that it is the distinctive mode of being of a "communicatively structured form of life."[82] Importantly, on this view, prenatal forms of life don't yet possess dignity because they are still incapable of engaging in reciprocal and symmetrical relations of mutual recognition with other human beings:

> As a member of a species, as a specimen of a community of procreation, the genetically individuated child *in utero* is by no means a fully-fledged person "from the very beginning." It takes entrance in the public sphere of a linguistic community for a natural creature to develop into both an individual and a person endowed with reason.[83]

The advantage of this strategy is that it allows Habermas to disconnect the debate on genetic intervention from the much too delicate ideological discussion of the moral status of human embryos. If he sees genetic intervention as a threat to human dignity, this is not because it tampers with prenatal forms of life but because it creates a radically asymmetric power balance between the generations, in the sense that the members of one generation would be treating those of the next as a means and not as ends, as objects and not as potential co-subjects. Hence, when Habermas argues that the danger of genetic technologies is that they have the power to change human nature, then he means that they will undermine the very foundations of the moral community of equals.

81 Habermas, *The Future of Human Nature*, 41-42.

82 *Ibid.*, 72.

83 *Ibid.*, 35.

Human beings possess dignity only insofar as they "belong to the universe of members who address intersubjectively accepted rules and orders to each other."[84] In other words, according to Habermas, notions such as 'dignity' and 'inviolability' "have a significance only in interpersonal relations of mutual respect, in the egalitarian dealings among persons."[85] As already said, this implies that the human in its presubjective state (embryo, fetus, newborn) cannot be given "the absolute protection of life enjoyed by persons who are subjects possessing basic rights."[86] Although these presubjective stages of development are a necessary precondition for the formation of one's own sense of personal identity, the human achieves moral agency and freedom precisely in its separation from this biological layer. Habermas also explains this distinction in terms of the difference between "being-a-body" and "having-a-body." The primary mode of self-experience is that of being a body. That is to say, in their initial stages of ontogenic development, human beings still fully coincide with their body, almost in the very same way as animals coincide with their environment. Having-a-body, on the other hand, "is the result of assuming an objectivating attitude toward the prior fact of being a body, a capacity we do not acquire until youth."[87] While initially we are one with our body, gradually we come to realize that this mode of experience does not fully exhaust our being and that we are also always more than just our body. Such an experience arises, for example, when we try to acquire a new skill and we are faced with the fact that our body just doesn't seem to want to cooperate with 'us.' Yet, a person "'has' or "possesses' her body only trough being this body in proceeding with her life."[88] It is only because we are our bodies that we are able to "differentiate between actions we ascribe to ourselves and actions we ascribe to others,"[89] and thus to understand ourselves as autonomous and unique persons. Moreover, Habermas adds, in order to feel one with our body, "it seems that that this body has to be experienced as something natural – as a continuation of the organic, self-regenerative life from which the person was born."[90] With genetic intervention, however, rather than being-a-body and having-a-body being distinct modes of experience, they enter a zone of indistinction:

---

[84] *Ibid.*, 33.

[85] *Ibid.*

[86] *Ibid.*, 42-43.

[87] *Ibid.*, 50.

[88] *Ibid.*

[89] *Ibid.*, 58.

[90] *Ibid.*

> The realization that our hereditary factors were, in a past before our past, subjected to programming, confronts us on an existential level, so to speak, with the expectation that we subordinate our being a body to our having a body.[91]

If a person realizes that her body has not been given her by nature, but that it has been genetically programmed by somebody else, then that person will experience the body that she 'is' permanently as the body that she 'has,' as thus as not quite her own. And insofar as this situation "may give rise to a novel, curiously asymmetrical type of relationship between person,"[92] it will prevent genetically modified humans and non-genetically modified humans from ever recognizing one another as moral equals.

At the end of the first chapter, we have argued that this line of argument also has a profoundly unsettling corollary. We argued that by giving substantive content to the notion of 'human nature,' Habermas's argument risks withholding the status of human being to any genetically enhanced being that may be brought into existence in the future. In his view, 'normal' human beings will not be able to enter into "symmetrical relationships" with these kinds of beings because they constitute "a foreign body in the legally institutionalized relations of recognition in modern societies."[93] Following Agamben and Esposito, we could say that by constructing his argument in this way, Habermas ends up establishing a biological caesura within humanity between genetically enhanced and non-genetically enhanced human beings. That is, by making membership of the moral community contingent upon the possession of a 'natural body,' Habermas risks isolating a "bare life" or an "existence without life" within the community.

It should be clear, then, that the problems and questions associated with the emerging possibility to radically redesign life through genetic technologies cannot be properly addressed within a human nature framework. Rather, what is urgently needed today is an equally radical politico-philosophical redefinition of life. Above all, this new conception of life should be conceived such that it no longer allow for a separation between pure biological life and political life, between *zoē* and *bios*. In chapter four we argued that Arendt's concept of natality provides us with the basic outlines of such a "form-of-life." If the potentiality of human beings to act freely is predicated on the condition that they are born biologically 'unfinished,' then this implies that the biological and political dimensions of human existence are radically

91 *Ibid.*, 54.

92 *Ibid.*, 42.

93 *Ibid.*, 14.

co-implicated from the very beginning. Moreover, we also suggested that this notion of natal life has strong implications for how we should approach the question of genetic intervention, for if it is true that the biological traits human beings are born with never solidify into a set of fixed properties and dispositions, then there is no reason to fear that genetic intervention will interfere with the human capacity to act. After all, even genetically enhanced human beings will still have to come into the world through birth.

Stiegler's theory of anthropogenesis provides two additional insights to this rearticulation of the concept of life. First, it allows us to see that human existence is not only natal on an ontogenetic level, but on an a phylogenetic level as well. We argued that the concept of natality expresses the idea that human freedom is conditioned by the biological fact that individual human beings are always born too early. By this we meant that our premature birth releases us from the fate of being compelled to follow a biogenetically predetermined course of life. Stiegler's theory argues that this is also true for the human species as a whole. To see this, we need only to briefly revisit his reading of the myth of Epimetheus and Prometheus. Remember that this myth states that the human only came into existence after Epimetheus had already handed out all available qualities to the other living beings and thus that the human was from the very beginning left unprovided for – "naked, unshod, unbedded, and unarmed."[94] So, in a sense, this myth also tells us that the human is 'born' both too early and too late. Too early because the human was brought into existence before it was endowed with a particular, defining quality; Too late because, at the time the human came into existence, all qualities had already been distributed to the other living beings. Yet, contrary to the standard interpretations of this myth[95] – and this is the second point we wish to make –, Stiegler doesn't take Prometheus's intervention to mean that the human is compensated for his originary lack of essence by the gift of technology. The human is born neither *before* technology (it is not an instrument), nor, for that matter, *after* technology but *through* technology. This is, at least, the conclusion that one may draw from Stiegler's reading of Leroi-Gourhan. As we explained, Stiegler argues that there is ample paleontological evidence to conclude that the process of hominization was the result of the fact that one particular living being managed to free itself from the pressure of genetic constraints by entering into a mutually constitutive relationship with technical artifacts. The birth of the human

[94] Quoted in Stiegler, *The Fault of Epimetheus*, 187.

[95] See for example Heinrich Popitz, *Der Aufbruch zur Artifiziellen Gesellschaft* (Tübingen: J. C. B. Mohr, 1995), 44-47.

is the transition from a genetic program to an epiphylogenetic program; or, what amounts to the same, from a process of genetic selection to a process of artificial selection. In a way, then, the human is always already technologically 'enhanced.' Clearly, this view has important consequences for how the contemporary debate on human genetic intervention is framed. Because if it is true that artificial selection has made the process of humanization possible in the first place, then it no longer makes any sense to be either 'for' or 'against' human genetic modification. As we've already indicated, seeing technology as an intrinsic condition of humanity doesn't in the least imply that genetic intervention is entirely unproblematic. But it does mean that it cannot be decided in advance what it essentially means to be human. Perhaps, then, the very possibility to genetically modify human beings retroactively reveals the fact that "we already were no longer human."[96]

[96] Stiegler, *The Fault of Epimetheus*, 136.

References

Agamben, G., *Language and Death: The Place of Negativity*, trans. K. E. Pinkus and M. Hardt (Minneapolis: The University of Minnesota Press, 1991).

Agamben, G., *Idea of Prose*, trans. M. Sullivan and S. Whitsitt (Albany: SUNY Press, 1995).

Agamben, G., *Homo Sacer: Sovereign Power and Bare Life*, trans. D. Heller-Roazen (Stanford: Stanford University press, 1998).

Agamben, G., *Means Without Ends: Notes on Politics*, trans. V. Binetti and C. Casarino (Minneapolis: University of Minnesota Press, 2000).

Agamben, G., *Remnants of Auschwitz: The Witness and the Archive*, trans. D. Heller-Roazen (New York: Zone Books, 2002).

Agamben, G., *The State of Exception*, trans. K. Attell (Chicago: The University of Chicago Press, 2005).

Agamben, G., *The Time that Remains: A Commentary on the Letter to the Romans*, trans. P. Dailey (Stanford: Stanford University Press, 2005).

Agar, Nicholas (1998), 'Liberal Eugenics,' *Public Affairs Quarterly*, 12 (2), 137-155.

Agar, N., *Liberal Eugenics: In Defence of Human Enhancement* (Malden, MA/ Oxford: Blackwell, 2004).

Annas, G. J. and Elias, S., *Gene Mapping: Using Law and Ethics as Guides* (Oxford: Oxford University Press, 1992).

Annas, G. J., *American Bioethics. Crossing Human Rights and Health Law Boundaries* (Oxford: Oxford University Press, 2005).

Arendt, H., *The Jew as Pariah* (New York: Grove Press, 1978).

Arendt, H., *Love and Saint Augustine* (Chicago: The University of Chicago Press, 1996).

Arendt, H., *The Human Condition* (Chicago: The University of Chicago Press, 1998).

Arendt, H., *Denktagebuch 1950-1973*, ed. U. Ludz and I. Nordmann (München/ Zürich: Piper, 2003).

Arendt, H., *The Origins of Totalitarianism* (New York: Schocken Books, 2004).

Arendt, H., *Between Past and Future. Eight Exercises in Political Thought* (New York: Penguin, 2006).

Beardsworth, R. (1995), 'From a Genealogy of Matter to a Politics of Memory: Stiegler's Thinking of Technics', *Tekhnema*, 2, 1-25.

Beardsworth, R., *Derrida & the Political* (London & New York: Routledge, 1996).

Beardsworth, R. (1998), 'Thinking Technicity', *Cultural Values*, 2 (1), 70-86.

Beiner, R., 'Action, Natality and Citizenship: Hannah Arendt's Concept of Freedom,' in *Conceptions of Liberty in Political Philosophy*, ed. Z. Pelczynski and J. Gray (London: Athlone Press, 1984).

Benhabib, S., *The Reluctant Modernism of Hannah Arendt* (Oxford: Rowan & Littlefield, 2003).

Bennington, G. (1996), 'Emergencies,' *Oxford Literary Review* 18, 175-216.

Binding, K. and Hoche, A., *Die Freigabe der Vernichtung Lebensunwerten Leben: Ihr Mass und ihre Form* (Leipzig: Meiner, 1920).

Birmingham, P., *Hannah Arendt and Human Rights. The Predicament of Common Responsibility* (Bloomington: Indiana University Press, 2006).

Blank, R. H. and Hines, S. M., *Biology and Political Science* (New York: Routledge, 2001).

Boorse, C. (1977), 'Health as a Theoretical Concept,' *Philosophy of Science* 44 (4): 542-573.

Bostrom, N. (2003), 'Human genetic Enhancements: A Transhumanist Perspective,' *Journal of Value Inquiry* 37 (4): 493-506.

Bowen-Moore, P., *Hannah Arendt's Philosophy of Natality* (Basingstoke/London: The Macmillan Press, 1989).

Bradley, A., 'Originary Technicity? Technology and Anthropology,' in *Technicity*, eds. A. Bradley and L. Armand (Prague: Litteraria Pragensia, 2006), 78-100.

Buchanan, A., Brock, D.W., Daniels, N., and Wikler, D., *From Chance to Choice: Genetics & Justice* (New York: Cambridge University Press, 2000).

Callarco, M., *Zoographies. The question of the animal from Heidegger to Derrida* (New York: Columbia University Press, 2008).

Campbell, A., Gillet, G. and Jones, D. G., *Medical Ethics*, 2nd edition (Oxford: Oxford University Press, 1999).

Canovan, M., *The Political Thought of Hannah Arendt* (London: Methuen, 1974).

Carter, N., *The Politics of the Environment: Ideas, Activism, Policy* (Cambridge: Cambridge University Press, 2001).

Cohen, S.N., Chang, A.C.Y., Boyer, H.W. and Helling, R.B. (1973), 'Construction of Biologically Functional Bacterial Plasmids *In Vitro*,' *Proc. Natl. Acad. Sci. USA* 70: 3240-3244.

Crick, F. (1970), 'The Central Dogma of Molecular Biology,' *Nature* 227: 561-563.

Dana, K. and Nathans, D. (1971), 'Specific cleavage of simian virus 40 DNA by restriction endonuclease of *Hemophilus influenzae*,' *Proc. Natl. Acad. Sci. USA* 68: 2913-2917.

Davies, J. C., *Human Nature in Politics: The Dynamics of Political Behavior* (New York: Wiley, 1963).

Derrida, J., *Of grammatology*, trans. G. C. Spivak (Baltimore: John Hopkins University Press, 1974).

Derrida, J., *Writing and difference*, trans A. Bass (Chicago: The University of Chicago Press, 1978).

Derrida, J. (1987), 'On reading Heidegger: an outline of remarks to the Essex colloquium,' *Research in Phenomenology* 17, 171-185.

Derrida, J., *Of spirit. Heidegger and the question*, trans. G. Bennington and R. Bowlby (Chicago: The University of Chicago Press, 1989).

Derrida, J., *Aporias*, trans. T. Dutoit (Stanford: Stanford University Press, 1993).

Derrida, J., 'Plato's Pharmacy,' in *Dissemination*, trans. B. Johnson (London & New York: Continuum, 2004), 69-186.

Derrida, J., *The animal that therefore I am*, trans. D. Wills (New York: Fordham University Press, 2008).

Derrida, Jacques, and Elisabeth Roudinesco. 2004. *For what tomorrow... A dialogue*, trans. Jeff Fort (Stanford: Stanford University Press).

Dupuy, J.-P., 'Cybernetics is Antihumanism: Advanced Technologies and the Rebellion Against the Human Condition,' in *Transhumanism and its Critics*, ed. G. R. Hansell and W. Grassie (Philadelphia: Metanexus, 2011), 227-248.

Durst, M. (2003), 'Birth and Natality in Hannah Arendt,' *Analecta Husserliana* Vol. 79, 777-797.

Dworkin, R., *Sovereign Virtue: The Theory and Practice of Equality* (Cambridge, MA: Harvard University Press, 2000).

Esposito, R., *Bíos: Biopolitics and Philosophy*, trans. T. Campbell (Minneapolis: The University of Minnesota Press, 2008).

Esposito, R., *Communitas: The Origin and Destiny of Community*, trans. T. Campbell (Stanford: Stanford University Press, 2010).

Esposito, R., *Immunitas: The Protection and Negation of Life*, trans. Z. Hanafi (Cambridge: Polity Press, 2011).

Esposito, R., *Community, Immunity, Biopolitics*, trans. R. N. Welch (New York: Fordham University Press, 2013).

Feinberg, J., 'The Child's Right to an Open Future,' in *Whose Child? Children's Rights, Parental Authority, and State Power*, ed. W. Aiken and H. LaFollette (Totowa, NJ: Rowan & Littlefield, 1980).

Feinberg, J. (1986), 'Wrongful life and the counterfactual element in harming,' *Social Philosophy and Policy* 4 (1): 145-178.

Flynn, B., *Political Philosophy at the Closure of Metaphysics* (Atlantic Highlands: Humanities Press, 1992).

Foucault, M., *Discipline and Punish: The Birth of the Prison* (New York: Pantheon Books, 1977).

Foucault, M., *The History of Sexuality, Vol. 1: The Will to Knowledge* (New York: Penguin Books, 1998).

Foucault, M., *Society Must Be Defended: Lectures at the Collège de France, 1975-76*, trans. D. Macey (London: Penguin Books, 2003).

Foucault, M. (2004), 'The crisis of Medicine or the crisis of antimedicine?' *Foucault Studies* 1: 5-19.

Foucault, M., *Security, Territory, Population: Lectures at the Collège de France 1977-1978*, trans. G. Burchell (Basingstoke: Macmillan, 2007).

Foucault, M., *The Birth of Biopolitics: Lectures at the Collège de France 1978-1979*, trans. G. Burchell (Basingstoke: Palgrave Macmillan, 2008).

Fox, D. (2007), 'The Illiberality of "liberal eugenics",' *Ratio* 10 (1): 1-25.

Fukuyama, F., *Our Posthuman Future. Consequences of the Biotechnological Revolution* (London: Profile Books, 2002).

Funke, O. (1985), 'Biopolitics and public policy: controlling biotechnology,' *Political Science* 18 (1): 69-77.

Galton, F., *Inquiries Into Human Faculty and its Development* (London: J. M. Dent & Company, 1883).

Gehlen, A., *Man: His Nature and Place in the World*, tr. C. McMillan and K. Pillemen (New York: Columbia University Press, 1988).

Glover, J., *What Sort of People Should There Be?* (New York: Penguin Books, 1984).

Habermas, J., 'Technology and Science as "Ideology",' in *Toward a Rational Society*, trans. J. J. Shapiro (London: Heinemann, 1971), 81-122.

Habermas, J., *Knowledge and Human Interests*, trans. J. J. Shapiro (London: Heinemann, 1972).

Habermas, J. (1973) 'A postscript to *Knowledge and Human Interests*,' *Philosophy of Social Science* 3: 157-189.

Habermas, J., *The Future of Human Nature* (Cambridge: Polity Press, 2003).

Habermas, J., 'An Argument against Human Cloning. Three Replies,' in *The Postnational Constellation. Political Essays*, trans. and ed. M. Pensky (Malden, Mass.: Polity Press, 2007).

Harris, J., *Enhancing Evolution: The Ethical Case for Making Better People* (Princeton: Princeton University Press, 2007).

Heidegger, M., *Being and Time*, trans. J. Macquarrie and E. Robinson (New York: Harper & Row, 1962).

Heidegger, M., *Discourse on Thinking*, trans. J. M. Anderson and E. H. Freund (New York: Harper Perrenial, 1966).

Heidegger, M., *Martin Heidegger im Gespräch*, ed. R. Wisser (Freiburg/München: Verlag Karl Alber, 1970).

Heidegger, M., *Basic Writings*, ed. D. F. Krell (New York: Harper & Row, 1977).

Heidegger, M., *The Question Concerning Technology*, trans. W. Lovitt (New York: Harper & Row, 1977).

Heidegger, M., *Heraklit* (Frankfurt a/M: Klostermann Verlag, 1979).

Heidegger, M., *Denkerfahrungen 1910-1976* (Frankfurt a/M: Klostermann Verlag, 1983).

Heidegger, M., *The Heidegger Controversy: A Critical Reader*, ed. R. Wolin (Cambridge Mass.: MIT Press, 1993).

Heidegger, M., *Bremer und Freiburger Vorträge* (Frankfurt a/M: Klostermann Verlag, 1994).

Heidegger, M., *Basic Questions of Philosophy: Selected Problems of Logic*, trans. R. Rojcewicz and A. Schuwer (Bloomington: Indiana University Press, 1994).

Heidegger, M., *The Fundamental Concepts of Metaphysics: World, Finitude, Solitude*, trans. W. McNeill and N. Walker (Bloomington: Indiana University Press, 1995).

Heidegger, M., *Pathmarks*, ed. W. McNeill (Cambridge: Cambridge University Press, 1998).

Heidegger, M., *Introduction to Metaphysics*, trans. G. Fried and R. Polt (New Haven: Yale University Press, 2000).

Heidegger, M., *Four Seminars*, trans. A. Mitchell and F. Raffoul (Bloomington: Indiana University Press, 2003).

Heidegger, M., *The End of Philosophy*, trans. and intro. J. Stambaugh (Chicago: The University of Chicago Press, 2003).

Heidegger, M. and Fink, E., *Heraclitus Seminar*, trans. C. H. Seibert (Evanston: Northwestern University Press, 1993).

Horkheimer, M. & Adorno, T.W., *Dialectic of Enlightenment*, trans. J. Cumming (New York: Herder & Herder, 1972).

Jacob, F., *The Logic of Life*, trans. Betty E. Stillmann (New York: Pantheon Books, 1974).

Jonas, H., *Philosophical Essays. From Ancient Creed to Technological Man* (Chicago and London: The University of Chicago Press, 1974).

Jay, M. (1978), 'The Political Existentialism of Hannah Arendt,' *Partisan Review* 45 (3): 348-369.

Jay, M., 'The Political Existentialism of Hannah Arendt,' in *Hannah Arendt: Critical Assessments of Leading Political Philosophers*, Vol. 3, ed. G. Williams (New York: Routledge, 2006), 191-213.

Jost, A., *Das Recht auf den Tod* (Göttingen: Grunow & Co., 1895).

Kass, L. (2003), 'Ageless Bodies, Happy Souls: Biotechnology and the Pursuit of Perfection,' *The New Atlantis* 1: 9-28.

Kevles, D. J., *In the Name of Eugenics. Genetics and the Uses of Human Heredity* (Cambridge MA: Harvard University Press, 1985).

Kitcher, P., *The Lives to Come: The Genetic Revolution and Human Possibilities* (London: Penguin Books, 1996).

Kjellén, R., *Staten som Lifsform* (Stockholm: Hugo Geber, 1916).

Kjellén, R., *Grundriss zu einen System der Politik* (Leipzig: Rudolf Leipzig Hirzel, 1920).

Kjellén, R., *Der Staat als Lebensform* (Berlin: Kurt Vowinckel Verlag, 1924).

Lawlor, L., *This is not sufficient. An essay on animality and human nature in Derrida* (New York: Columbia University Press, 2007).

Lemke, T., *Biopolitics: An Advanced Introduction*, trans. E. F. Trump (New York and London: New York University Press, 2011).

Leroi-Gourhan, A., *Gesture and Speech*, trans. A. B. Berger (Cambridge Mass.: MIT Press, 1993).

Lesch, K., Greenberg, B. D., Higley, J. D., *et al.*, 'Serotonin transporter, personality, and behavior,' in *Molecular Genetics and the Human Personality*, eds. J. Benjamin, R. Ebstein and R. Belmaker (Washington, DC: American Psychiatric Publishing, 2002).

Marcuse, H., *Eros and Civilization* (Boston: Beacon Press, 1955).

Marcuse, H., *One-Dimensional Man* (Boston: Beacon Press, 1964).

Masters, R. D., *The Nature of Politics* (New Haven: Yale University Press, 1989).

McCarthy, T., *The Critical Theory of Jürgen Habermas* (Cambridge Mass.: MIT Press, 1981).

Mckibben, B., *Enough! Staying Human in an Engineered Age* (New York: Henry Holt & Company L. L. C., 2003).

McLaren, A., 'Prospects for pre-implantation diagnosis of genetic disease,' in B. Holland and C. Kyriacou (eds.), *Genetics and Society* (Wokingham: Addison-Wesley, 1993), 67-85.

Mendieta, E. (2003), 'Communicative freedom and genetic engineering,' *Logos* 2 (1): 124-139.

Mill, J. S., *On liberty* (Cambridge: Cambridge University Press, 1989).

Mills, C., *Futures of Reproduction. Bioethics and Biopolitics* (Dordrecht: Springer, 2011).

Moravec, H., *Robot: Mere Machine to Transcendent Mind* (Oxford: Oxford University Press, 1998).

Nozick, R., *Anarchy, State and Utopia* (New York: Basic Books, 1974).

O'Byrne, A., Natality *and* Finitude (Bloomington: Indiana University Press, 2010).

Parfit, D., *Reasons and Persons* (Oxford: Clarendon Press, 1984).

Passerin d'Entrèves, M., *The Political Philosophy of Hannah Arendt* (London: Routledge, 2003).

Paul, D.P. (1992), 'Eugenic Anxieties, Social Realities, and Political Choices,' *Social Research* 59 (3): 663-683.

Pitkin, H., *The Attack of the Blob. Hannah Arendt's Concept of the Social* (Chicago: The University of Chicago Press, 2000).

Ploetz, A., *Die Tüchtigkeit unserer Rasse und der Schutz der Schwachen* (Berlin: Fischer Verlag, 1895).

Popitz, H., *Der Aufbruch zur Artifiziellen Gesellschaft* (Tübingen: J. C. B. Mohr, 1995).

Portmann, A. (1941), 'Die Tragzeiten der Primaten und die Dauer der Schwangerschaft beim Menschen: ein Problem der vergleichen Biologie,' *Revue Suisse de Zoologie*, Vol. 48, 511-518.

Rager, G. (2008), 'Is preimplantation genetic diagnosis ethically acceptable?' *Bioethica Forum* 1 (2): 81-88.

Rancière, J. (2004), 'Who is the subject of human rights?' *South Atlantic Quarterly* 103: 297-310.

Reiter, H., 'Unsere Biopolitik und das Auslandsdeutschtum,' in *Das Reichsgesundheitsamt 1933-1939: Sechs Jahre Nationalsozialistische Führung* (Berlin: Julius Springer Verlag, 1939).

Reiter, H. and Breger, J. (eds.), *Deutsches Gold: Gesundes Leben – Frohes Schaffen* (München: Röhrig, 1942).

Roberts, B. (2005), 'Stiegler Reading Derrida: The Prosthesis of Deconstruction in Technics,' *Postmodern Culture*, 16 (1).

Roberts, B. (2006), 'Rousseau, Stiegler and the Aporia of Origin,' *Forum for Modern Language Studies*, 42 (4), 382-394.

Robertson, J. A., *Children of Choice: Freedom and the New Reproductive Technologies* (Princeton: Princeton University Press, 1994).

Rojcewicz, R., *The Gods and Technology: A reading of Heidegger* (Albany: SUNY Press, 2006).

Rousseau, J. J., 'A discourse on the Origin of Inequality,' in *Great Books of the Western World, Vol. 38: Montesquieu and Rousseau*, ed. R. M. Hutchins (Chicago: Chicago University Press, 1952).

Sandel, M. J., *The Case Against Perfection. Ethics in the Age of Genetic Engineering* (Cambridge MA: The Belknap Press of Harvard University Press, 2008).

Savulescu, J. (2001), 'Procreative beneficence: why we should select the best children,' *Bioethics* 15 (5/6): 413-426.

Savulescu, J. (2009), 'The moral obligation to create children with the best chance of the best life,' *Bioethics* 23 (5): 274-290.

Schmitt, C., *Political Theology. Four Chapters on the Concept of Sovereignty*, trans. G. Schwab (Chicago: The University of Chicago Press, 2005).

Scully, J. L. (2001), 'Drawing a Line: Situating Moral Boundaries in Genetic Medicine,' *Bioethics* 15 (3): 189-204.

Silver, L.M., *Remaking Eden: How Genetic Engineering and Cloning Will Transform the American Family* (New York: Avon Books, 1998).

Somit, A. and Peterson, S. A., *Darwinism, Dominance, and Democracy: The Biological Bases of Authoritarianism* (Westport: Praeger, 1997).

Stiegler, B., 'Quand faire, c'est dire. De la technique comme différance de toute frontière,' in *Le passage des frontières. Autour du travail de Jacques Derrida*, ed. Colloque de Cérisy (Paris: Galilée, 1994).

Stiegler, B., *La Technique et le Temps 1. La Faute d'Épiméthée* (Paris: Galilée, 1994).

Stiegler, B., *La Technique et le Temps 2. La Désorientation* (Paris: Galilée, 1996).

Stiegler, B., *La Technique et le Temps 3. Le Temps du Cinéma et la Question du Mal-être* (Paris: Galilée, 2001).

Stiegler, B., *Technics and Time 1. The Fault of Epimetheus*, trans. R. Beardsworth & G. Collins (Stanford: Stanford University Press, 1998).

Stiegler, B. (2003), 'Technics of decision,' *Angelaki* 8 (2), 151-168.

Stiegler, B., *Passer à l'Acte* (Paris: Galilée, 2003).

Stiegler, B., *Philosopher par Accident* (Paris: Galilée, 2004).

Stiegler, B., 'Anamnēsis and Hypomnēsis: The Memories of Desire,' in *Technicity*, eds. A. Bradley & L. Armand (Prague: Litteraria Pragensia, 2006).

Stock, G., *Redesigning Humans: Our Inevitable Genetic Future* (Boston/ New York: Mariner Books, 2002).

Tang, Y., Shimzu, E., Dube, G. R., Rampon, *et al.* (1999), 'Genetic enhancement of learning and memory in mice,' *Nature* 401: 63-69.

von Uexküll, J., *Staatsbiologie: Anatomie, Physiologie, Pathologie des Staates* (Berlin: Verlag von Gebrüder Paetel, 1920).

Van Camp, N. (2009), 'Stiegler, Habermas, and the techno-logical condition of man,' *Journal for Cultural Research* 13 (2), 125-141.

Van Camp, N. (2011), 'Negotiating the Anthropological Limit: Derrida, Stiegler, and the Question of the Animal,' *Between the Species*, 14 (1), 57-80.

Van Camp, N. (2012), 'Heidegger and the question concerning biotechnology,' *Journal of Philosophy of Life* 2 (1), 32-54.

Vanhagen, T., *Prospects of Democracy: A Study of 172 Countries* (New York: Routledge, 1997).

Vogel, S., *Against Nature: The Concept of Nature in Critical Theory* (Albany: SUNY Press, 1996).

von Verschuer, O., *Leitfaden der Rassenhygiene* (Leipzig: Thieme, 1941).

Walters, L. and Palmer, J. G., *The ethics of Human Gene Therapy* (New York: Oxford University Press, 1997).

Werber, N. (2002), 'Die Normalisierung des Ausnahmefalls: Giorgio Agamben sieht immer und überall Konzentrationslager,' *Merkur* 56: 618-622.

Whitebook, J. (1979), 'The Problem of Nature in Habermas,' *Telos* 40, 41-69.

Wilkinson, S. (2008), '"Eugenics talk" and the language of bioethics,' *Journal of Medical Ethics* 34: 467-471.

Wolin, S. (1983), 'Democracy & The Political,' *Salmagundi* 60: 3-19.

Wolin, R., *Heidegger's Children. Hannah Arendt, Karl Löwith, Hans Jonas, and Herbert Marcuse* (Princeton, NJ: Princeton University Press, 2001).

Young-Bruehl, E., *Hannah Arendt. For Love of the World* (New Haven/London: Yale University Press, 1982).

Žižek, S., *On Belief* (New York/London: Routledge, 2001).

Žižek, S., 'Of Cells and Selves,' in *The Žižek reader*, eds. Elizabeth Wright and Edmond Wright (Malden Mass.: Blackwell, 2004), 302-320.

Žižek, S., *Organs Without Bodies* (New York/London: Routledge, 2004).

Žižek, S., *The Parallax View* (Cambridge Mass.: MIT Press, 2006).

"Philosophy & Politics"

Open to thinkers from all countries and cultures who are not limiting their efforts to understand the processes of social institutions, but are raising questions about their goals and meaning, this series hopes to contribute to a renewal of political philosophy. Beyond the different disciplines of social sciences, ideological commitments and the limits of "national schools" philosophy is bound to pursue its universal enquiry on human wisdom in a world chaotically pushed towards greater unity.

Series directed by ***Gabriel Fragnière***
and ***Hendrik Opdebeeck***

Series Titles

Confiance réflexive et institutionnalisme. Des théories libérales du choix rationnel à la gouvernance du fait social monétaire, Benjamin Six, n° 25, 2013, 226 p., ISBN 978-2-87574-092-2

A Study in Transborder Ethics: Justice, Citizenship, Civility, Geneviève Souillac, No. 24, 2012, 221 p., ISBN 978-90-5201-824-9

Creative Crises of Democracy, Joris Gijsenbergh, Saskia Hollander, Tim Houwen & Wim de Jong (eds.), No. 23, 2012, 444 p., ISBN 978-90-5201-797-6

Philosophical Essays. Critic Rationalism as Historical-objective Transcendentalism, Giulio Preti, edited by Fabio Minazzi, No. 22, 2011, 326 p., ISBN 978-90-5201-778-5

Regards éthiques sur l'Union européenne, Ignace Berten, Gabriel Fragnière, Philippe D. Grosjean, Peter Knauer, Daniel Spoel et Frank Turner, n° 21, 2011, 251 p., ISBN 978-90-5201-737-2

Responsibility and Cultures of the World. Dialogue around a Collective Challenge, Edith Sizoo (ed.), No. 20, 2010, 243 p., ISBN 978-90-5201-670-2

Between Enlightenment and Disaster. Dimensions of the Political Use of Knowledge, Linda Sangolt (ed.), No. 19, 2010, 171 p., ISBN 978-90-5201-631-3

Variations sur la confiance. Concepts et enjeux au sein des théories de la gouvernance, Claire Lobet-Maris, Robin Lucas & Benjamin Six (dir.), No. 18, 2009, 206 p., ISBN 978-90-5201-525-5

La cosmodémocratie. Un principe de gouvernance pour la société technologique et mondialisée, René Ngambele Nsasay, No. 17, 2008, 310 p., ISBN 978-90-5201-463-0

Un droit, des morales. Valoriser l'État laïque, Paul LÖWENTHAL, No. 16, 2008, 156 p., ISBN 978-90-5201-442-5

Robert Schuman: Neo Scholastic Humanism and the Reunification of Europe, Alan FIMISTER, No. 15, 2008, 284 p., ISBN 978-90-5201-439-5

Science and Ethics. The Axiological Contexts of Science, Evandro AGAZZI & Fabio MINAZZI (eds.), No. 14, 2008, ISBN 978-90-5201-426-5

Rôles, action sociale et vie subjective. Recherches à partir de la phénoménologie de Michel Henry, Raphaël GÉLY, n° 13, 2007, 205 p., ISBN 978-90-5201-347-3

Identités et monde commun. Psychologie sociale, philosophie, société, Raphaël GÉLY, n° 12, 2006 (3e tirage 2008), 206 p., ISBN 978-90-5201-416-6

Feminists Contest Politics and Philosophy, Lisa N. GURLEY, Claudia LEEB & Anna Aloisia MOSER (eds.), No. 11, 2005, 272 p., ISBN 978-90-5201-252-0

The Importance of Ideals. Debating Their Relevance in Law, Morality, and Politics, Wibren VAN DER BURG & Sanne TAEKEMA (eds.), No. 10, 2004, 274 p., ISBN 978-90-5201-226-1

Social Sciences and Political Change. Promoting Innovative Research in Post-Socialist Countries, Robin CASSLING & Gabriel FRAGNIÈRE (eds.), No. 9, 2003, 297 p., ISBN 978-90-5201-168-4

Working-Class Women in Elite Academia. A Philosophical Inquiry, Claudia LEEB, No. 8, 2004, 220 p., ISBN 978-90-5201-979-6

L'autonomie éthique. Débat démocratique et vérité, Giuseppe G. NASTRI, n° 7, 2002, 200 p., ISBN 978-90-5201-972-7

Plus est en l'homme. Le personnalisme vécu comme humanisme radical, Vincent TRIEST, n° 6, 2000 (4e tirage 2004), 214 p., ISBN 978-90-5201-922-2

Enlightenment and Genocide, Contradictions of Modernity, James KAYE & Bo STRÅTH (eds.), No. 5, 2000, 278 p., ISBN 978-90-5201-919-2

Les défis du nationalisme moderne. Québec, Catalogne, écosse, Michael KEATING, n° 4, 1997, 298 p., ISBN 978-90-5201-705-1

www.ingramcontent.com/pod-product-compliance
Lightning Source LLC
Chambersburg PA
CBHW071343230926
51846CB00002B/429
9782875742810